# WHY INDIANS HAVE NOT INVENTED ANYTHING

*How a Civilization Chose*

*the Inner Universe Over the Outer*

RAJENDRA PRADHAN

First Edition, 2026

Published by Hormones River

Printed in India

ISBN: 978-93-5361-425-6

www.whyindians.com

## A note on the contents

This book contains references to ancient Indian texts, philosophical concepts, and traditional practices. Sanskrit and Hindi terms are translated into English in brackets at first use within each chapter. Where the book refers to scientific research, it does so in general terms and as one of several ways of understanding the topics discussed.

This book is not intended as medical, financial, legal, or psychological advice. It is a work of philosophical and cultural inquiry. Readers should consult appropriate professionals for guidance in any of these areas.

## A short glossary of recurring terms

Ten words return throughout this book. Each is translated at its first appearance in every chapter, but the reader may find it useful to meet them once, together, before beginning.

Raag — desire, attachment, longing; identified by Ayurveda as the root disease of human suffering.

Purnam — wholeness, completeness; lacking nothing.

Dravya — material wealth; literally "that which flows," from the root for liquid.

Mahalakshmi — prosperity in its complete form, of which money is only one part.

Buddhi — clear intelligence, wisdom; always established before wealth.

Prana — the life force, the body's inner current.

Dharma — right action; the order that sustains life.

Seva — service, offered without keeping account.

Ahankar — the ego born from knowledge and accomplishment.

Vasudhaiva Kutumbakam — "the world is one family," the root claim of the entire civilization.

## Author contact and feedback

The author welcomes thoughtful feedback from readers. Contact details are available at www.whyindians.com.

# Contents

**ॐ पूर्णमदः पूर्णमिदं**

**पूर्णात् पूर्णमुदच्यते।**

**पूर्णस्य पूर्णमादाय**

**पूर्णमेवावशिष्यते॥**

**ॐ शान्ति शान्ति शान्ति**

*— Isha Upanishad, Shanti Mantra*

*That is complete. This is complete.*

*From completeness, completeness emerges.*

*Even when completeness is taken from completeness,*

*only completeness remains.*

*Peace. Peace. Peace.*

वह पूर्ण है, यह भी पूर्ण है। पूर्ण से ही पूर्ण उत्पन्न होता है ।

पूर्ण में से पूर्ण निकाल लेने पर भी पूर्ण ही शेष रहता है ।

## The Question

A child is born.

In the first weeks of life, that child needs nothing from the world. The mother's milk is enough. There is no thirst that needs water from outside, no hunger that needs solid food, no question the child is asking that has not already been answered by simply being alive. The newborn is whole. Complete. Lacking nothing.

This is the most common miracle on earth. It happens millions of times a day, in every corner of the world, with so little ceremony that almost nobody stops to consider what is actually happening. A new human being arrives, and for a brief period, that human being is in a state of completeness that they will spend the rest of their life trying, in various unconscious ways, to return to.

Then the child begins to eat.

And with food — with the first cry for what is outside, the first reaching for what is not the mother — desire begins. The journey of an entire human lifetime begins. The journey of every civilization that human beings have ever built begins.

This book is, in one sense, about that moment. About what happens before it, what happens after it, and about a civilization that spent five thousand years studying it more carefully than any other civilization on earth.

---

I have been an observer for as long as I can remember.

Watching what my grandmother did in the kitchen without explaining why. Listening to what the family said at births and at funerals. Noticing which rituals my mother performed at certain times of the year, and which ones she would never perform at others. Paying attention to the small gestures — the way an older relative would touch the head of a younger one before a long journey, the specific way an Indian woman wraps a sari, the exact moment in a meal when the elders begin to eat. None of these things were explained to me. None of them were taught. They were simply there, woven into ordinary life, carrying knowledge that nobody could quite put into words.

I did not know I was observing. I only knew that India, in its daily life, was carrying something that no one was speaking aloud.

The question came later. It arrived one evening, on my birthday. I was alone. I had left a job some time earlier and had nothing waiting for me on the other side — no offer, no plan, only the feeling that something else was supposed to happen. To pass the evening, I put on a film I had seen before.

Interstellar. A film about humanity searching for a new home as the earth slowly fails.

There is a small scene early in the film. The main character — an American father — is driving through a cornfield with his two children. A drone appears, drifting slowly above the field, no operator in sight. The children chase it, laughing. The father catches it eventually and sees what it is: an Indian surveillance drone that has wandered, somehow, all the way across the world. The father looks at the drone with something closer to respect. He tells them: this thing has been flying on solar power for years. He keeps the solar cells.

It is not a big scene. The film moves on to other things. But that night, sitting alone, something in that scene caught me and would not let go. The American children laughing as they chased it. The father — the wiser one in the scene — recognizing what was actually impressive about the Indian invention.

And a question rose up in me. Not from anywhere in particular. Just — there.

*Why have Indians not invented anything?*

I knew, even as the question formed, that it was not a fair question. India had given the world zero, the decimal system, the foundations of mathematics without which no modern technology could exist. The civilization that produced Ayurveda, the Vedas, the Upanishads, the Bhagavad Gita. A civilization whose contribution to human thought is so vast that the question itself was, in some sense, an insult.

But the question was not asking what India had given. The question was asking something more uncomfortable. Why is it that, in the modern era — the age of the steam engine, the printing press, the telephone, the airplane, the antibiotic, the computer, the internet — Indian names appear so rarely on the list of inventors? Why does a civilization with five thousand years of accumulated knowledge appear to have stepped aside, just at the moment when invention became the measure of civilizational worth?

That night the question entered me. It did not leave.

But the question, I came to understand later, was not really new. The question was everything I had been observing since childhood, finally gathered into words. The grandmother in the kitchen, the woman wrapping the sari, the specific moment the elders began to eat — all of it had been the same question, in a quieter form. What is this knowledge that Indian life carries, that nobody speaks aloud, that the modern measuring instruments cannot detect?

For more than a decade, in one form or another, I have carried it. Through job changes and life changes. Through reading and conversation. Through hundreds of small encounters with the question in different forms — in books, in arguments, in the casual remarks of foreigners who assumed certain things about India and in the casual remarks of Indians who had absorbed those same assumptions about themselves.

I asked the question of many people. I asked it of educated professionals and of villagers who had never finished school. I asked it of family members and of strangers on trains. The answers were many, and I will share them in their proper place in this book.

But none of the answers ever satisfied me. Because every one of them assumed the same thing — that not inventing was a failure. That India was, somehow, missing something.

This book does not begin with an answer, or even with a better assumption. It begins with an exploration. But before I take you into it, you should know who is offering it.

---

I am from Rajasthan, the desert state in India's northwest — a land where water is scarce, and where you learn early that what is scarce must flow carefully, and what flows carefully sustains everything. I grew up close to Pushkar — the city where, according to the ancient texts, Brahma the creator performed the first yajna, the first sacred fire, to bring this universe into being. And when hardship moved me, as a boy, into the town's government school — the one everyone said had a bad image — I was moved, without knowing it, into the most important classroom of my life. What it taught me waits deep inside this book, where it belongs. I grew up among ordinary Indian people, in an ordinary Indian family, with ordinary Indian struggles. There is nothing in my biography that would qualify me to write this book.

Except that the question came to me. And the observation that gave rise to it had been there from the beginning. And it stayed.

What I have written here is not the work of a scholar. I do not have the academic credentials that this kind of book is normally supposed to have. I have read what I could, observed what I could, asked questions of people willing to answer, and sat for long hours with what I was reading and what I was hearing.

What I am offering is not a history. It is not a defence. It is not, despite the provocative title, an argument that Indians are superior to anyone or that any other civilization is inferior. The title is a question that the world has been asking for two centuries. This book is what one Indian man, sitting with that question across many years of careful observation, came to understand about it.

What I came to understand is that the question contains a hidden assumption. The assumption is that invention — the building of new machines, new systems, new technologies — is the measure of a civilization's worth. Once you accept this assumption, India looks like a civilization that fell behind. Once you question it, India looks like a civilization that chose differently. And once you understand what India was choosing instead, the entire question reorganizes itself.

India was not failing to invent. India was inventing something different. Something that does not show up in archaeological digs or patent records. Something that the modern measuring instruments are not calibrated to detect.

What that something is — and why it may turn out to be more important, in the age that is now beginning, than anything the modern world has invented — is the subject of this book.

---

This book is not an answer. It is a remembering.

If you are a reader from outside India, I invite you to read it as you might read a thoughtful traveller's account of a country you have visited briefly. There will be things that surprise you. There will be claims you want to push back on. That is good. The book is not asking you to agree. It is asking you to consider whether the question you have been asked about India was the right question.

If you are a reader from India, I invite you to read it slowly. Some of what is here you will already know. Some of it you will recognize even if you have never heard it said in this way. And some of it — the most important parts — you will know because you carry it in your body, in your kitchen, in the practices your grandmother passed to your mother and your mother passed to you, even if no one ever told you what those practices were really for.

The shloka that appears on the page before this preface is the oldest framing of what this book is about. It is from the Isha Upanishad. You may read its translation and still not know what it means. That is fine. By the end of the book, you will.

**You already know everything in this book. You simply forgot.**

**Let us remember together.**

*— Raj*

*Rajasthan, India*

CHAPTER ONE

# Raag

*The Disease Called Desire*

रागादिरोगान् सततानुषक्तानशेषकायप्रसृतानशेषान् ।

औत्सुक्यमोहारतिदाञ्जघान योऽपूर्ववैद्याय नमोऽस्तु तस्मै ॥

— *Ashtanga Hridayam (Heart of the Eight Branches of Medicine), Verse 2*

*Salutations to the unique Physician who destroyed, without any residue, all diseases like Raag (desire/attachment) — which are constantly associated with the body, giving rise to disease, delusion and restlessness.*

## The Question Nobody Asked

The world has a question about India. It is not a new question. It has been asked in university lecture halls, in the pages of history books, in late-night conversations between economists and philosophers, and in the quiet, uncomfortable thoughts of Indians themselves.

The question is this: Why have Indians not invented anything?

No steam engine. No printing press. No telephone. No airplane. No penicillin. No theory of relativity. For a civilization that is among the oldest on this planet — a civilization that produced extraordinary mathematics, philosophy, literature, and art — the absence of modern invention is, to many observers, a puzzle that demands an explanation.

I have heard many answers to this question. I have sat with Indians in living rooms, in chai (tea) shops, in offices, and on the floors of temples, and asked them. Their answers have been as varied as the people themselves. Some said we were lazy. Some said we lacked resources. Some said colonization stole our spirit. Some shrugged, as if the question itself was an insult they had long since learned to endure.

But none of these answers ever satisfied me. Because every one of them assumed the same thing — that not inventing was a failure. That the absence of the steam engine was a gap. That India was, somehow, missing something.

This book begins with a different question. What if India was not missing anything?

What if India was not missing anything? What if the absence of invention was not a failure, but a choice? And what if that choice was rooted in the most sophisticated understanding of human nature that any civilization has ever produced?

**To understand that choice, we must begin with Raag.**

## The Disease at the Root of Everything

More than fifteen hundred years ago, a physician named Vagbhata sat down to write the most important medical text in Indian history. He called it the Ashtanga Hridayam (the Heart of the Eight Branches of Medicine). This text became one of the three foundational texts of Ayurveda (India's ancient science of life and health). It is a work of extraordinary scientific and philosophical depth.

And the very first thing Vagbhata chose to write — before anatomy, before surgery, before any discussion of herbs or treatments — was a salutation to the physician who had destroyed the diseases called Raag completely, without residue.

The greatest medical text in Indian history does not begin by talking about fever, or infection, or broken bones. It begins by identifying Raag — desire, attachment, longing — as the root disease of the human body.

This is India's most fundamental statement about the nature of human suffering. And, in many ways, modern medicine has arrived at a related understanding. What India called Raag shares much with what modern medicine calls stress. A growing body of research suggests that chronic stress elevates cortisol (the stress hormone), drives systemic inflammation, suppresses immune function, and is implicated in much of what we now call psychosomatic illness. The World Health Organization has recognized stress as one of the major health concerns of our time.

Vagbhata wrote about this in the fifth century. Modern medicine reached related conclusions fifteen hundred years later.

*The root cause of disease — physical, mental, and spiritual — is desire. And if desire is the root disease, then every invention in human history is, in some sense, a symptom.*

## What Health Actually Means

Before we examine the faces of Raag (desire), it is worth understanding what Ayurveda (ancient Indian health science) considered the opposite of disease — because it is not merely the absence of illness.

Ayurveda defines a healthy being as Svasthah (one who is established in the self). And the condition of this healthy being is described as Prasannatmendriyamanah — body, mind, and senses in a state of harmony and satisfaction.

This is not a medical definition. It is a civilizational one. Ayurveda is not saying that health means no fever. It is saying that health means your body, mind, and senses are in complete harmony — the state that exists when Raag is absent.

**The modern world invents medicines to remove symptoms. India built a civilization to restore Svasthah — the healthy state of being established in oneself, the condition from which symptoms do not arise in the first place.**

## The Nine Faces of Raag

Raag is not a simple word. It is a vast family of inner states — each one a different face of the same fundamental disease. Indian philosophy identifies nine primary expressions. And when you examine them carefully, you will see something extraordinary: every significant invention in human history traces back to at least one of them.

Kama (desire for pleasure) — every invention for comfort was born from this. The air conditioner, the soft mattress, the heated car seat. The body wants to feel good. And so it invents.

Krodha (anger / desire to dominate) — every weapon ever created. From the first sharpened stone to the nuclear bomb. The need to overpower, to win, to destroy what threatens you.

Lobha (greed) — every financial system, every mechanism for accumulating wealth beyond what is needed. The feeling that what you have is never enough.

Moha (attachment / delusion) — every invention designed to preserve what we love. To protect, to store, to extend the life of things and people. The desperate human need to hold on.

Mada (pride / arrogance) — every invention that exists primarily to impress others. The faster car, the taller building, the more powerful nation. The deeply human need to be seen as better.

Matsarya (jealousy / comparison) — so much invention happens not because something is needed, but because someone else has it first. Nations build weapons not because they need them, but because others have them.

Asmita (identity-ego) — we are the nation that put a man on the moon. We are the company that changed the world. So much of invention is the invention of a story about ourselves.

Abhinivesha (fear of death) — medicine, surgery, anti-aging research, life extension technology. All of it traces back to the primal terror of ending. The refusal to accept that the body is temporary.

Ahankar (ego born from knowledge) — the most subtle and perhaps most dangerous form. The specific disease of the educated and accomplished. The need to prove that your intelligence is superior. That the contribution of your mind will outlast everything else. In the age of technology, it is the engine of almost everything.

Is there a single significant invention in human history that does not trace back to at least one of these nine forces? I have thought about this for many years. I have tried to find the exception. I have not found it.

**Every invention is a child of Raag.**

## The Endless Cycle

But Raag does not simply drive invention. It drives an endless cycle — a loop from which, once entered, it is almost impossible to escape.

A human being feels Raag. They want something they do not have. They suffer from its absence. So they invent something to satisfy that desire. For a moment they feel satisfied. But the satisfaction does not last. Because the nature of desire is not to be satisfied. The nature of desire is to shift. To expand. To find a new object, a new lack, a new suffering.

We invented the car because we did not want to walk. The car gave us pollution, traffic, and obesity from inactivity. So we invented cleaner fuels, traffic management systems, and treadmills. Each solution generating a new problem. Each new problem generating a new invention. On and on, without end.

We invented the mobile phone to communicate more easily. The mobile phone gave us addiction, anxiety, broken relationships, and an inability to be present in our own lives. So we invented digital detox programs, screen time limits, and meditation apps to help us escape the phone we could not put down.

We invented antibiotics to defeat infection. Antibiotics gave us antibiotic-resistant superbugs. We invented social media to connect people. Social media gave us the loneliness epidemic.

And now — in the age of artificial intelligence — this cycle is accelerating beyond anything previous generations experienced. Everyone can see what everyone else is building, achieving, accumulating. Matsarya (jealousy) and Ahankar (ego) have found a new engine of extraordinary power. Everyone is in a hurry. Everyone feels late. Everyone is churning furiously, trying to produce something before someone else produces it first.

And yet, in the middle of this acceleration, India still keeps living counter-examples on ordinary street corners. Walk into a traditional mithai (sweets) shop in any old Indian town. Many of them make only a handful of items — the same few sweets their grandfathers made, fresh each morning, finished by evening. Now walk into a modern shop with two hundred varieties in a refrigerated glass wall. Ask which shopkeeper is at peace. The one with five items wakes up knowing exactly what today holds. The one with two hundred begins every morning forecasting — which sweets to make more of, which less, what will sell, what will pile up unsold by night. He has more choice, more inventory, more customers — and he has surrendered the present moment to a permanent meeting with the future. The traditional shopkeeper did not lack imagination. He had made a calculation the modern world has forgotten how to make: that every option added to the display is a thought added to the mind. More variety, more Raag. Fewer things, done completely — and the mind stays where the hands are.

The pattern was understood with complete clarity thousands of years before the first car, the first phone, or the first algorithm. It was understood in the opening verse of the Ashtanga Hridayam.

**The disease is Raag. The treatment is not more invention. The treatment is the physician who removes Raag itself — from the root, without residue.**

## A Necessary Clarity

**Before we go further, I want to be clear about something. This book does not say that invention is wrong.**

The wheel, the written word, medicine, music, the technologies that connect human beings across continents — these are among the most beautiful expressions of human creativity. India did not reject invention. India asked a deeper question first:

*Why do you want to invent? And what will it cost?*

When invention comes from wisdom — from genuine understanding of what life needs, what the ecosystem can sustain, what the human body and mind truly require — it nourishes. When it comes purely from Raag — from fear, from greed, from the desperate need to prove oneself superior — it carries the seeds of its own destruction.

The question is not whether to invent. The question is from where the invention comes. This distinction — between conscious invention and Raag-driven invention — is the central argument of this book.

## The Physician Who Removes the Root

Who is the physician that Vagbhata salutes in the opening verse of the Ashtanga Hridayam? He is Dhanwantari (the divine physician of Indian tradition) — the one who carries the pot of Amrita (immortal nectar). He did not appear from nowhere. He emerged from the Samudra Manthan (churning of the cosmic ocean) — which we will explore fully in Chapter Four.

But the tradition also holds something more intimate than this cosmic origin. Dhanwantari (the divine healer) does not only exist outside. He can be awakened within. Because if every human being is born with the complete knowledge of the universe — as this book will argue — then the physician who removes Raag is also within. Waiting for the right conditions to emerge.

**Nobody else can cure your Raag. No doctor, no guru, no algorithm. Only the Dhanwantari within you — awakened through consciousness, through stillness, through the daily practice of choosing what this book will describe — can remove the root disease.**

## When Knowledge Becomes Raag

There is something I have observed about Indian people that I want to share with you — not as a criticism, but as one of the most honest observations I have ever made about my own culture.

**Every Indian gives health advice.**

You mention a headache, and three people immediately tell you what to do. Nobody has a medical degree. But everybody has an answer.

Where does this confidence come from? It comes from something real. India built a civilization where healing knowledge was so widely distributed — so deeply encoded into daily rituals, seasonal practices, fasting traditions, and the rhythms of family life — that ordinary people carried extraordinary wisdom about the human body. Not in books. In their habits. In practices passed from mother to daughter across hundreds of generations.

This was Ayurveda in its living form. Not confined to textbooks or clinics, but alive in the kitchen, in the temple, in the daily routine of every Indian home.

But something happened. Ayurveda was gradually replaced in hospitals and in the minds of educated Indians by an entirely different system. And yet the confidence remained.

**Confidence without the wisdom behind it is the most sophisticated form of ignorance.**

This is Ahankar in its most modern and common form. The container survived. The contents changed. And nobody noticed. India's greatest strength — a civilization-wide distribution of healing wisdom — became, in its disconnected form, a quiet vulnerability.

## India's Answer to Raag

So what was India's answer? If Raag is the disease — the root cause of all suffering, all restlessness, all invention driven by lack — what was the treatment India prescribed for itself?

The answer was not silence, or passivity, or the rejection of life. The answer was something far more radical and far more sophisticated.

India built its entire civilization around one project: the removal of Raag from the human experience. Not by force. Not by law. Not by suppression. But by understanding. By daily practice. By encoding the treatment into every aspect of life — into the clothes people wore, the food they ate, the rituals they observed, the stories they told their children.

Every chapter of this book is one piece of that project. A different dimension of how India chose to address the root rather than the symptom.

Before we go there, I want to leave you with a question. A question that I believe is the most important question this book asks.

If every invention is a child of Raag — and if Raag is a disease — then what does that make the world we have built?

**And what would the world look like if, just once, a civilization chose to treat the disease instead of celebrating the symptom?**

India chose that. For thousands of years.

**And the world called it — falling behind.**

CHAPTER TWO

# White

*The Ancient Science of Equality*

## The Room

**Imagine a room full of the most accomplished people in the world.**

Scholars. Scientists. Philosophers. Physicians. Each one has spent a lifetime building something extraordinary — a body of knowledge, a field of mastery, a reputation earned through decades of work. They have gathered in this room to discuss their ideas, to share their discoveries, to debate what they know.

And very quickly, the discussion becomes something else.

Voices rise. Arguments sharpen. Each person leans forward with the particular intensity of someone who needs to be right. The room fills with the invisible electricity of competing egos, each one pushing against the others.

This is simply what happens when accomplished human beings gather together. The ego — Ahankar (ego born from knowledge), as we explored in the previous chapter — does not disappear when a person becomes knowledgeable. It grows. It finds more sophisticated reasons to assert itself. It wears the costume of intellectual debate while carrying, underneath, the oldest human need:

*I matter. I am significant. I am more than you.*

Now — in the middle of all this heat and noise — imagine something very simple happens. Someone looks around. Really looks. And notices — everyone is wearing white.

The greatest scholar in the room — white. The youngest student — white. The most celebrated physician — white. The man who has never published a single word — white. Simple, plain, unadorned white cloth on every body in the room.

And something shifts. Not dramatically. Just quietly, like a tide going out. The argument loses some of its urgency. Because the eyes have delivered a message that the mind cannot easily ignore:

*We are the same. We are wearing the same thing. So why exactly are we fighting?*

This is not an accident. This is ancient technology — working exactly as it was designed to work.

India's tradition imagined this same scene at a far higher altitude — and with far higher stakes.

Picture not a conference of scholars but a gathering of Gurus — the kind the tradition speaks of taking place in the silence of the Himalayas, far from ordinary eyes. Beings who have spent lifetimes in the disciplines this book will describe in Chapter Seven, and who carry, in the tradition's understanding, powers beyond anything a degree can certify. Their capabilities are not equal. They were never equal. Each has walked a different path to a different depth. And if beings of such power were to turn against one another, the tradition understood, the consequences would not be an unpleasant meeting. They would be measured on the earth itself.

Now let them look at one another. Every body in that gathering is wrapped in the same simple white cloth. The one whose word could move mountains and the one still in his first years of practice — the same white. And in that single glance, before any contest can begin, the thought completes itself: we are equal — not in power, but in what we are. The fight ends before it starts. Not because anyone was defeated. Because the cloth removed the question.

This is the deepest function of white, and the reason the most powerful needed it most. Among ordinary people, clothing hides small differences of wealth and rank. Among the realized, it holds together beings whose real differences are invisible, unequal, and immense — by keeping one truth in front of the eyes at all times: whatever the powers, the wearer is the same.

## The Ego's Greatest Weapon

The ego has many weapons. But its first and most immediate weapon — the one it uses before words are spoken, before character is revealed, before any real information about a person has had a chance to register — is appearance.

Before a word is spoken, before a thought is shared, before any quality of a person's character or intelligence or kindness has had a chance to express itself

— the eyes have already made their judgment. They have looked at the clothes. The colors. The fabric. The brand. The cut. The jewelry. The watch. The shoes.

And within seconds — a hierarchy has been established. This person is above me. This person is below me. This person is worth my full attention. This person can be dismissed.

This is Raag (desire/attachment) operating through the eyes. This is Mada (pride) and Matsarya (jealousy/comparison) doing what they always do. Dividing. Ranking. Separating human beings from each other based on what they have chosen to wrap around their bodies.

India looked at this mechanism — this ancient, powerful, deeply human tendency to judge and rank and separate through appearance — and made a choice.

**Remove the weapon.**

If the ego uses appearance to divide — then remove the difference in appearance. Wear the same thing. Wear the simplest thing. Wear white.

Not as a religious obligation. Not as a cultural tradition mindlessly followed. But as a daily, conscious, deliberate act of removing Raag (desire) from the very first thing another human being sees when they look at you.

This is what the kurta (traditional shirt) and dhoti (traditional lower garment) represent. This is what the simple white cloth of the Indian tradition was always doing — quietly, every single day, on the bodies of millions of people who may not have known the philosophy behind what they wore, but were living it nonetheless.

## The Science Behind the Color

The choice of white was not only philosophical. It was scientific. Multiple dimensions of science — operating simultaneously in one simple color.

Thermal science first. White reflects sunlight rather than absorbing it. In the intense heat of the Indian subcontinent — where temperatures regularly exceed forty degrees Celsius — wearing white keeps the body significantly cooler than any other color. The simple loose cut of the traditional kurta (shirt) and dhoti (lower garment) allows air to circulate freely around the body, reducing heat buildup. This was not fashion. This was engineering — clothing designed for the specific climate of the land where it was worn.

Color psychology next. Research on color and behavior has consistently observed that visual uniformity reduces the brain's automatic hierarchy-building activity. When the eyes have less information to work with — when one person looks like another in terms of clothing — the mind has fewer raw materials to assemble status comparisons from. The ego's first line of defence is weakened. This is not an absolute claim about color psychology, which remains a contested scientific field. But it aligns with what observation has long suggested about how human beings respond to visual sameness.

And spiritually — white is the combination of all colors. Not the absence of color. The presence of all of them simultaneously, unified into one. In that sense, white clothing on an Indian sage or scholar carried a profound message:

*I contain everything. Yet I show nothing. I am complete — and I have nothing to prove.*

This is Purnam (wholeness/completeness) — the principle from the shloka (verse) at the beginning of this book — expressed through clothing. Complete. Whole. Nothing missing. Nothing needing to be displayed.

## Color Has Its Own Vocabulary

If white was the color of equality — the daily reduction of Raag — Indian civilization also understood that color, used deliberately and in the right context, carried its own precise meaning. White was not the absence of color. It was the resting state. The condition to which the system returned. Other colors were tools — used carefully, for specific purposes, at specific times.

Walk into any Indian home during a major puja (worship ritual) or anushthan (sacred ceremony) and observe what has been laid down on the floor or the platform where the ritual will take place. The cloth spread there is rarely white. The color has been chosen with care, by people who in many cases could not articulate the reasoning, but whose practice carried the reasoning intact across thousands of years.

Red — the color of Shakti (feminine creative energy), of vital force, of the active principle. A red linen is spread for ceremonies invoking the goddess in her warrior and creative forms — Durga (the warrior goddess), Lakshmi (the goddess of prosperity), Kali (the fierce protective form). Red is also the color of vermilion — the sindoor that a married Indian woman wears in her hair parting, signaling the active life-giving force she carries. Red is what the system uses when energy is to be invoked rather than dissolved.

Yellow — the color of Saraswati (the goddess of wisdom and learning), of turmeric, of the auspicious beginning. Yellow appears at weddings, at ceremonies of new beginnings, at rituals where knowledge and clarity are being requested. The pre-wedding haldi ceremony — when turmeric paste is applied to the bride and groom — is a yellow ritual. The morning puja for Ganpati (Lord Ganesh, the remover of obstacles) often uses yellow flowers and yellow cloth. Yellow signals: clear the channel. Make the path bright. Begin clean.

Black — counter-intuitively, used in specific protective rituals and in the worship of certain forms of Shiva (the lord of transformation) and Kali (the fierce protective goddess). Black absorbs rather than reflects. In the Indian symbolic system, black is the color that contains all colors at the absorbing level — as white contains all colors at the radiating level. A black thread tied around a child's wrist is not superstition; it is the encoded recognition that black absorbs the energetic noise the child should not be carrying.

Green for Buddh (Mercury) and growth. Orange and saffron for the renunciate and the ascetic — the colors of fire and discipline. Each color has been mapped, over thousands of years of observation, to a specific energetic effect that the practitioner is inviting into the ritual space.

**Color was a vocabulary. Like Sanskrit, like mantra, like mudra — color was a precise language for communicating with the energetic structure of reality.**

And white, in this larger context, becomes even more meaningful. White is not the lack of vocabulary. White is the silence between the words. The neutral state from which all other colors emerge and to which they return. Wearing white in daily life was the practitioner's way of saying: I do not need to invoke. I do not need to absorb. I am at rest, in the resting color, in the natural state of the system. Today I am simply being.

## Vasudhaiva Kutumbakam — The Deeper Root

The philosophy of white clothing is not an isolated practice. It is rooted in something much larger that runs through the entire Indian civilizational understanding:

*Vasudhaiva Kutumbakam (the world is one family).*

From the Maha Upanishad (one of the ancient Indian philosophical texts), this phrase is not merely a sentiment of universal brotherhood. It is a statement about equality that goes to the root of why white works. If the world is one family — if every human being is a member of the same household — then dressing to signal superiority over any member of that family is a contradiction in terms.

You do not compete with your family in the way you dress. You do not try to signal your superiority to your own mother or your own child. Within the family, the ego's weapons of appearance lose their power — because everyone already knows each other's value.

White clothing was India's daily, embodied expression of this understanding. Every morning, putting on the white kurta was an act of saying: you are my family. I do not need to impress you. I do not need to prove myself to you. We are the same.

## The World Discovered the Same Thing

India was not alone in this discovery. Across many cultures and many centuries, the most conscious human beings — independently, without knowing what India had understood — arrived at a similar conclusion: simplify your appearance, and something in the mind becomes clearer.

Walk into any professional office in the modern world. Look at what people are wearing. White shirts. Dark trousers. The same thing, more or less, on every body. It is the same ancient instinct that India encoded into the kurta and dhoti — the desire to create a level playing field, where a person is judged by what they say and do rather than what they wear.

And the principle is not only historical. Travel today through the Gulf — the United Arab Emirates, Saudi Arabia, the wider Arab world — and observe what men wear. The same simple white garment, on the wealthy and the modest alike, every day, as a matter of cultural identity. Whatever its own history and reasons, the effect is the one this chapter has been describing: in a room where everyone wears white, the eyes are given nothing to rank. An entire modern society, one of the most prosperous on earth, still dresses daily in the color of visual equality — a living demonstration, on another continent, of the principle India encoded into the kurta and dhoti.

The professional suit attempted the same thing. A uniform for the modern working world. But the suit failed where the Indian white succeeded. Because Raag found the gaps and filled them. The brand of the suit. The quality of the fabric. The watch on the wrist. The shoes on the feet.

**The ego found the gaps and filled them.**

India's white was more complete. A simple white kurta is a simple white kurta. There is very little room for Mada to hide inside it.

And perhaps the most striking modern echoes of the principle come from the world of technology and high-pressure leadership, where a number of well-known figures have, in recent decades, made a deliberate practice of wearing the same simple clothing every day. None of them, as far as is publicly known, did this for spiritual or philosophical reasons. They did it as a productivity choice — a way of removing the daily decision of what to wear, of making appearance irrelevant, of freeing mental energy for what actually mattered.

It is a small thing. A personal efficiency choice in each case. But it is also, from the longer Indian view, a rediscovery — by a few individuals, in the most pressured corners of modern life — of a principle that an entire civilization had encoded into daily practice thousands of years earlier.

**Simplify the outside. And the inside becomes more powerful.**

The difference is this: where modern productivity advice tells one person to wear the same shirt to save mental energy, India built a civilization where everyone, every day, wore the same thing as a daily medicine against Raag, embedded into the fabric — quite literally — of everyday life.

## The Daily Reminder

Think about what it means to put on white clothing every morning. Not once. Not on a special occasion. Every single day. As the first conscious act of the morning — before food, before work, before conversation — you dress your body in the color of completeness. The color that says: I contain everything, and I need to display nothing.

This is not a philosophy you read in a book and then forget. This is a philosophy that touches your skin every morning. That you carry on your body throughout the day. That every mirror and every other human being reflects back to you, moment after moment:

*You are the same as me. We are the same. There is nothing to prove here.*

This is how India encoded the treatment for Raag into daily life. Not through lectures or laws or religious commandments. Through the simplest possible act: choosing what to wear in the morning.

And this principle extended far beyond clothing. It extended to homes — simple, functional, unadorned. To food — prepared with attention to health rather than display. To speech — measured, considered, without the need to impress. To the entire way a life was constructed and lived.

**A civilization that does not need to show off has no reason to invent things to show off with.**

The luxury car exists because Mada exists — the need to be seen as successful. The designer label exists because Matsarya (jealousy) exists — the need to be seen as equal to or better than the person next to you. The mansion exists because Ahankar (ego from knowledge and achievement) exists — the need for your success to be visible to others.

Remove Raag. And the market for all of these things — not the things themselves, but the Raag-driven version of them — simply disappears.

## White as Prophecy

There is one more dimension of white that must be mentioned here. And it is perhaps the most profound.

India did not choose white only because it was practical. India chose white because it could see where humanity was going.

Think about the greatest conversations happening in the world right now. Across science, philosophy, technology, ethics — the most thoughtful people of every culture are arriving, slowly and from very different starting points, at remarkably similar conclusions. About interconnection. About the nature of consciousness. About the limits of materialism. About the cost of unchecked Raag. About what it means to live well.

The conversation is converging. Different vocabularies. Same underlying recognition. Different traditions. Same destination.

India saw this. Five thousand years ago. India saw that — given enough time, given enough suffering, given enough cycles of invention and consequence

— humanity would arrive at the recognition that the differences between us were always smaller than the sameness beneath them.

**And India chose to dress for that destination, in the present moment, every single day.**

White was not the starting point. White was the destination. India just got there first.

## What India Forgot

And then India forgot.

Walk through any modern Indian city and look at what people are wearing. Branded shirts. Designer labels. Expensive watches. The same status competition that the Western world spent the twentieth century perfecting — now reproduced, with enthusiasm, on Indian bodies.

White is no longer the default. White, in many urban Indian contexts, has become the color of waiters and household staff and people of insufficient status to wear something more interesting. White has been demoted. The thing that was once a daily practice of Purnam (wholeness) is now, in too many places, a marker of being less.

This did not happen by accident. Colonization left its physical structures behind — the railways, the administrative systems, the offices — but its deepest legacy was a colonization of the imagination. A generation of educated Indians learned to associate Indian dress with backwardness and Western dress with progress. The white kurta became something to escape from. The Western suit and shirt became something to aspire to.

India forgot the science. India forgot that the simple white cloth was carrying knowledge — about the body, about the ego, about the deepest social purpose of clothing. India forgot that wearing white was not poverty; it was sovereignty. Sovereignty over the ego, over the need for approval, over the daily war of appearances.

**To remember white is to remember Purnam.**

## An Invitation

This chapter is not a call to abandon all color. It is not an instruction to dress in any particular way. It is an invitation to remember the principle that white was carrying.

Tomorrow morning, when you open your wardrobe and choose what to wear, pause for one second. Notice what the choice is actually about. Is this clothing chosen for comfort? For the practical demands of the day? Or is it being chosen, even partly, to communicate something to others — to signal status, to invite approval, to differentiate yourself from those who will see you?

There is no wrong answer. The point is not to suppress the impulse. The point is to see it clearly. To notice the ego's quiet operation in something as ordinary as dressing. To begin to understand that Raag operates in a hundred small ways every day, in choices we did not even know we were making.

The deeper invitation is this: somewhere in the daily life that you can shape, find one place to introduce the principle of white. One meal eaten without performing it for anyone. One conversation conducted without trying to impress. One day, one room, one ordinary corner of life kept simple — not because simplicity is virtuous, but because the absence of Raag is the natural state, and every small return to that state strengthens the receiver that this book will, in its later chapters, ask you to consider.

**White was India's daily medicine. The principle behind it is universal. The cloth itself is optional.**

In the next chapter, we move from clothing to wealth — and to a goddess who has been misunderstood for many centuries. We will look at Mahalakshmi (the goddess of complete prosperity) and at the precise, sophisticated economic philosophy India encoded into a single annual festival of light.

CHAPTER THREE

# Mahalakshmi

*The Physics of Prosperity*

## The Wrong Question

There is a question that has followed India for two centuries, asked by economists, journalists, and well-meaning visitors who could not reconcile what they saw.

**Why is a civilization with so much wisdom so poor?**

The question is wrong from the first word. But to see why, we have to start by understanding what India actually meant by wealth — because the Sanskrit (ancient Indian language) and Hindi vocabulary for wealth contains a precision that the modern English word "money" does not have.

In the modern world, wealth is a number. It is what your bank tells you you have. It is the digit on your tax return, the figure on your house valuation, the balance on your investment statement. Modern wealth is, almost entirely, what India would have called Dravya (liquid material) — the flowing, transactional, measurable form of value.

India had a different word for the complete picture. Mahalakshmi (the great Lakshmi, the goddess of complete prosperity). Mahalakshmi is what wealth looks like when you stop measuring it only with money.

## Two Forms of Lakshmi

The Indian tradition is clear that Lakshmi (the goddess of prosperity) has more than one form. The two that matter for this chapter are Mahalakshmi (the complete form) and Dravya Lakshmi (the material form). They are not in competition. They are not opposites. They are nested — Dravya Lakshmi sits inside Mahalakshmi as a small specific case sits inside a much larger whole.

Mahalakshmi is described in the Vedic texts as having three dimensions: Sampatti (prosperity in its complete form), Siddhi (achievement and accomplishment), and Nidhi (sacred treasure). She is the wealth that is visible and the wealth that is invisible. She is the wealth that is movable and the wealth that is immovable. She is the abundance of food in a house, the depth of knowledge in a mind, the warmth of relationships in a family, the peace of a settled heart, the auspiciousness of a home where guests are welcomed, the strength of a community where children grow up safely.

None of this can be put into a bank account. None of this shows up on a balance sheet. But India considered all of it Mahalakshmi (complete prosperity). The bank balance was Dravya Lakshmi — one specific liquid form, valuable but secondary.

**The modern world measures Dravya Lakshmi and concludes that India is poor. India measures Mahalakshmi and recognizes itself as one of the wealthiest civilizations on earth.**

These are two different questions producing two different answers. Neither is wrong by its own standard. They are simply not the same question.

## Padmaja — She Who Lives in the Lotus

One of Mahalakshmi's oldest names is Padmaja (she who is born of the lotus). She is shown, in every traditional depiction, either standing or seated on a fully opened lotus flower. She holds lotuses in her hands. Her crown is decorated with lotus motifs. The temples dedicated to her are surrounded by lotus ponds. Wherever Mahalakshmi appears in Indian art, the lotus appears with her.

This is not decoration. The lotus is doing precise philosophical work.

Consider what kind of flower the lotus is. It does not grow in clean water or in fertile soil. It grows in mud — in pond bottoms, in shallow swamps, in the silt and decay of stagnant water. The mud is its only home. Without the mud, there is no lotus. And yet — the lotus does not absorb the mud. It rises through the water. It opens into the air. The petals come up dry, immaculate, separated from the substrate that gave them birth. A lotus flower in full bloom is one of the most strikingly pure objects in nature, and it was just yesterday surrounded by the slime that grew it.

**The lotus is the visible proof that something completely pure can emerge from a source that is completely impure — if it knows how to rise.**

This is what Padmaja (the lotus-born) is teaching about prosperity. Genuine wealth — the complete Mahalakshmi form — is not born in pristine conditions. It is born in the ordinary mud of daily life: in the difficulties of running a household, in the small frictions of commerce, in the inheritance of family struggles, in the practical compromises every working person has to make. The mud is where wealth begins. The question is whether the wealth knows how to rise through the water, or whether it stays stuck in the mud, taking on its color.

Wealth that stays in the mud is what the modern world calls dirty money. The same compromises, the same exploitations, the same accumulations driven by Raag (desire) — visible everywhere, in every economy, in every century. Wealth that rises through the water and opens above it is Padmaja's wealth. Same origin. Different relationship to the mud.

There is a second meaning to the lotus that is even more important for this book. The Puranas (ancient encyclopedic texts) describe that at the beginning of the universe, while Lord Vishnu (the preserver) lay in cosmic sleep on the serpent of infinity, a lotus emerged from his navel. The lotus opened. And out of that lotus, Lord Brahma (the creator) was born — and from Brahma came everything else.

The universe itself, in the Indian telling, was born in a lotus. The seat of creation was a flower that knew how to rise. This is not casual mythology. This is a precise statement about how creation operates: from a centered source, through the medium of water (which Chapter Six and Chapter Eight will return to), an opening that does not contaminate itself with the substrate beneath it.

**Mahalakshmi sits on the lotus. Brahma was born from the lotus. Saraswati (the goddess of wisdom) holds the lotus. Ganpati (the remover of obstacles) is seated on the lotus. The lotus is not the seat of any one deity. The lotus is the seat of the divine itself — the position from which whatever is most pure in existence emerges.**

When an Indian sits in front of a Lakshmi image on the night of Diwali (festival of lights) and places a marigold or a fresh lotus before her, the meaning is not religious decoration. The practitioner is being shown, in flower and image, the model: wealth that rises through the mud without taking on its color. Prosperity that does not require purity of source — only purity of trajectory.

**Ganpati First. Always.**

There is a detail in Indian ritual that is easy to miss and almost impossible to overstate the importance of. Whenever Lakshmi is worshipped — whether at Diwali, at the opening of a new business, at the housewarming of a new home, at any moment when prosperity is being invited — she is not worshipped first.

**Ganpati (Lord Ganesh, the remover of obstacles) is worshipped first. Always. Without exception.**

In the sacred texts, this precedence is so absolute that it extends even to the marriage of Shiva (the lord of transformation) and Parvati (his consort, the divine feminine). Their wedding is one of the foundational events in Hindu cosmology — the union of pure consciousness with pure energy that, in some traditions, sets the world into motion. And the pandit (priest) conducting that ceremony — the priest officiating at the union of the two most fundamental cosmic principles — began by invoking and honouring Ganpati.

Think about what this means. Shiva — one of the Trimurti (holy trinity of Brahma, Vishnu, Mahesh), one of the most powerful forces in the universe — could not begin his own marriage without the blessing of Ganpati. Ganpati is the son of Shiva and Parvati — and yet he precedes their union. He exists before the event that brings his own existence into being.

Ganpati represents Buddhi (pure intelligence) and Sat-Buddhi (truthful uncorrupted awareness). He is the clearing of the channel through which all creation flows. Before any genuine creation can occur — even the most cosmic and powerful — the channel must first be cleared of whatever obscures it.

Now back to Diwali. When Indians worship Mahalakshmi (goddess of complete prosperity) on Diwali, they always worship Ganpati first. Ganpati is placed on the right side — the position of precedence. The ancient texts are clear:

*Bhagwan Shri Ganesh is the master of Siddhi (achievement), Buddhi (intelligence), Shubh (auspiciousness) and Labh (gain). He destroys all Vighna (obstacles) and all inauspiciousness. He is the provider of Sat-Buddhi (pure truthful intelligence).*

The message encoded in this practice is one of the most practical pieces of financial wisdom ever given to humanity:

*Before wealth can flow — wisdom must be established. Buddhi before money. Understanding before accumulation. Always. Even for the most powerful forces in the universe.*

The modern world placed Dravya Lakshmi (material wealth) at the center of the educational system, the career system, the social system. It teaches children mathematics for money, science for industry, language for commerce. And it wonders why so many intelligent people make terrible decisions with wealth. Why so much accumulated money produces so little genuine wellbeing.

India's answer: you put Lakshmi before Ganpati. You placed Dravya (material wealth) before Buddhi (wisdom). And when you do that — the wealth that arrives does not bring Kalyan (wellbeing), Mangal (auspiciousness) and Anand (joy) that the texts promise. It brings only more Raag.

**Ganpati first. Always. Without exception.**

## The Physics of Dravya

Now we arrive at something extraordinary — one of the most original economic insights that Indian philosophy ever produced.

The word Dravya does not simply mean money. In Sanskrit, Dravya comes from the root Drava, which means liquid. That which flows. That which is fluid.

**India did not just name money. India described its fundamental nature.**

Money is liquid. And like all liquids, it obeys the most fundamental law of fluid dynamics:

*It flows downward. It follows the slope. It moves toward the lowest point.*

The lowest point in human terms is not the poorest or the weakest. It is the lowest in terms of ego. The most humble. The most grounded. The person who has released Raag — who does not clutch, who does not hoard, who does not fight the natural flow of things.

**Money, like water, flows toward the humble.**

## The Hormones River

The principle that India encoded into the word Dravya does not stop at money. It reaches deeper — into the body itself.

Consider what is flowing inside you right now. Beneath the skin, beneath the muscles, beneath conscious awareness — your bloodstream carries a continuous river of chemical messengers. Hormones. Insulin and glucagon controlling your blood sugar. Cortisol responding to stress. Adrenaline and noradrenaline managing your arousal and alertness. Oxytocin produced when you embrace someone you love. Serotonin and dopamine regulating your mood. Thyroid hormones setting the pace of your metabolism. Growth hormone repairing your tissues at night. Sex hormones across the long arcs of life. Dozens of these messengers, each a different signal, each in continuous motion through the same medium.

The hormonal system is, in the most literal sense, a river. Chemical messengers dissolved in liquid, carried by the heart's pump through arteries and capillaries to every organ, every tissue, every cell. The hormones do not stay where they are made. They flow. And like all flowing systems, they obey the same physics that India observed in Dravya (liquid wealth).

**The hormones flow toward the body that is open to receive them.**

Stress shuts the receivers. When the body is gripped by Raag — by the racing thoughts, the contracted muscles, the breath held too high in the chest — the cells become harder to reach. Cortisol rises and stays risen, becoming a continuous low-grade interference rather than a clean signal for a specific threat. Insulin resistance develops: the cells stop hearing the message. The hormonal river is still flowing, but the cellular doors that should open to its messages are jammed shut by the ego's constant alarm.

Calm opens the receivers. When the body is settled, the breath low and even, the mind released from grasping — the cells respond. Oxytocin reaches the heart and the heart slows. Serotonin reaches the gut and the gut relaxes. Growth hormone is released during deep sleep and the body repairs itself. The river still flows the same way; what has changed is the body's relationship to the river.

This is why every Ayurvedic (ancient Indian health science) practice of stillness — the morning routine of Dinacharya (daily regimen), the calming techniques of Pranayama (breath regulation), the prescribed silence of the early morning hour — was understood as a treatment, not a luxury. They were the technologies that opened the cellular doors so that the body's own internal pharmacy could deliver its medicine. The river was always there. The question was whether the body was quiet enough to receive what the river was bringing.

**Hormones cannot be hoarded. Money cannot be hoarded. Dravya (any liquid medium) cannot be hoarded. The attempt to hoard a liquid is the attempt to defy its own nature, and the attempt always fails.**

The body that tries to manufacture and accumulate its own cortisol — chronic stress — gets sick. The person who tries to accumulate Dravya (money) beyond their preparation — chronic Raag — gets poor in every dimension that money does not measure. The same physics. The same Dravya. The same warning, written into the human body and the human economy in the same ink.

India saw this. India named it once, in a single word, and let the word do the work of teaching whoever was willing to listen for the next three thousand years.

## The Dam — Why Individual Wealth Without Collective Preparation Fails

Here is an analogy that extends the physics of Dravya even further.

If you want to collect the water of a river — you must build a dam. Not a weak one. A powerful one. Because the dam must hold the weight of everything the river brings. A dam is not easy to build. It requires extraordinary preparation, extraordinary resources, and extraordinary collective effort. And once built, it requires constant maintenance.

Now apply this to Dravya Lakshmi. If you want to hold significant wealth — wealth beyond your basic needs, wealth that accumulates and grows — you need a dam strong enough to hold it. And that dam is not built by one person alone.

The dam is your entire family. Your community. Your civilization. If everyone around you thinks with clarity, lives with reduced Raag, serves genuinely, and understands the difference between Mahalakshmi and Dravya Lakshmi — then the dam is concrete. It holds. The water accumulates.

But if one person accumulates wealth while the family and community around them remains unprepared — the dam has cracks. The water leaks. And you cannot hold river water in a cracked container, no matter how much flows in.

*You cannot hold Dravya (wealth) beyond the strength of your preparation. And your preparation is not just yours — it is your family's, your community's, your civilization's. This is why individual prosperity without collective wisdom always eventually collapses.*

India understood this. Its approach to prosperity was always collective. The family first. Then the community. Then the civilization. Sat Chit Anand (truth-consciousness-bliss) — the state of true wellbeing — is only possible when the entire family experiences it together. Individual happiness built on the suffering of those around you is not Mahalakshmi. It is an illusion of Dravya Lakshmi that will not hold.

## The City and the Billions

Let me give you an analogy that has stopped people in their tracks every single time I have used it.

Imagine a city. A real, living, breathing city — with shops and restaurants and offices and markets and millions of daily transactions. In this city, on any given day, billions of dollars of value are circulating. Moving from hand to hand, account to account, business to business.

That billion is flowing constantly, like a great river through the landscape of the city.

**Within that flowing river — who receives more?**

Is it the person who grasps most desperately? Or is it the person who serves best — who listens, who adapts, who makes every interaction feel like a genuine exchange rather than an extraction?

Anyone who has run a real business over a long period of time will tell you the same answer. The shopkeeper who treats every customer with genuine respect keeps customers for life. The business that serves rather than extracts builds loyalty that no marketing campaign can buy. Money flows toward the person who has released ego from the transaction — whose Raag is quieter. Lower. More like the valley than the mountain.

**Water does not flow to the mountain. It flows to the valley.**

**Dravya does not flow to the ego. It flows to the humble.**

This is not mysticism, and it is not literal physics. It is a patient observation about how value actually moves between human beings — an observation India compressed into the physics-metaphor of a single word: Dravya (liquid/flowing material).

## What the World's Wealthiest People Accidentally Discovered

The world's most successful people have, in many cases, accidentally rediscovered half of what India understood about Dravya.

They discovered that arrogance repels. That genuine service attracts. That the leader who listens builds better companies than the one who only commands. The pattern shows up repeatedly in the biographies of long-term wealth holders: they live more modestly than their bank accounts allow, they describe their philosophy in terms of patience and stillness rather than aggression, they do not consume status as enthusiastically as their net worth would permit. Some of the most celebrated long-term investors of the modern era have continued to live in the same modest houses they bought decades ago, describing their approach to wealth in language that — stripped of its market vocabulary — sounds remarkably like the language of an Indian sage describing non-grasping.

**But here is where the modern wealth-holder and the Indian understanding part ways.**

The world's most successful people discovered that ego must be removed from external behavior — from how they deal with others, how they build teams, how they treat customers. But they kept the inner Raag. The insatiable desire for more. The need to accumulate beyond any rational necessity.

**They removed the external ego. But kept the internal Raag.**

India went one step further. India understood that true prosperity — Mahalakshmi in her complete form — requires the removal of both. The external ego AND the internal desire.

**This is the difference between building an empire and building a civilization.**

## Two Operating Systems for Human Civilization

India's ancient economy was not built on money. It was built on exchange — the direct trading of goods and services between people who needed what the other had.

The farmer gave grain. The potter gave vessels. The vaidya (healer) gave medicine. The guru (teacher) gave knowledge. Each person contributed what they had in abundance and received what they needed. The exchange was immediate, visible, human — a direct connection between two people, two needs, two offerings.

Now compare this to the money-based economy. The farmer sells grain for money. The money is spent on something entirely unrelated to grain. The value of the farmer's contribution is reduced to a number — a number that can be accumulated, hoarded, manipulated, inflated, deflated, gambled with.

The barter economy was a system of Mahalakshmi. Complete, balanced, human, direct. A system where the invisible wealth of relationship, trust, and community was as real and as valuable as any physical commodity.

The money economy is a system of Dravya Lakshmi (liquid material wealth). Powerful — but by its very nature, prone to accumulate at the lowest ego point, to concentrate in fewer and fewer hands, to separate the rich from the poor with a mathematical precision that no barter economy ever achieved.

India insisted that Dravya Lakshmi must always remain secondary to Mahalakshmi. That money must always serve life — not replace it.

## The World's Most Sophisticated Annual Financial Philosophy

Every year, on Kartik Krishna Amavasya (the darkest night of the month of Kartik) — Diwali — hundreds of millions of people light lamps, clean their homes, and perform a ritual that most of them would describe simply as worshipping the goddess of wealth.

But look at what the ritual actually instructs:

First — clean the home. Purify the space. Remove the old, the stale, the accumulated clutter. This is the removal of stagnant Raag and stagnant attachment to things that no longer serve.

Second — purify yourself. Before you can invite prosperity, you must be a clean vessel. A person full of Raag, ego, anger, attachment cannot receive Mahalakshmi's (complete prosperity's) full blessing.

Third — place Ganpati (Ganesha/lord of beginnings) first. Establish Buddhi before wealth. Light the lamp of intelligence before inviting the flow of money.

Fourth — draw the Ashtadal Kamal (eight-petalled lotus) in saffron and sandalwood paste. Place the physical money on this sacred geometry. Money is real, money has a place — but it must be held within a sacred structure of wisdom and cosmic order.

Fifth — worship both together. Acknowledge both Mahalakshmi and Dravya Lakshmi. Do not pretend money does not exist. Do not pretend it is everything. Hold both in right relationship.

This five-step ritual is performed by hundreds of millions of people every year. Most could not articulate its economic philosophy. But they live it — encoded into their annual practice, preserved across generations, transmitted not through books but through the simple act of doing what their parents did.

**This is India's knowledge preservation system at work. Not in universities. Not in textbooks. In the annual ritual of cleaning your home and lighting a lamp.**

## The Invitation of Diwali

### Every year, on the darkest night, India lights a billion lamps.

Not to celebrate money. Not to worship accumulation. But to invite the complete form of prosperity — Mahalakshmi in all her dimensions. The visible and invisible. The movable and immovable. The wealth that can be counted and the wealth that can only be felt.

To do this, India cleans first. Purifies first. Establishes Ganpati (wisdom/Ganesha) first. And only then invites Lakshmi to enter.

*Clean first. Purify first. Establish Buddhi first. And then — let Dravya flow where it naturally flows. Toward the humble. Toward the clean. Toward the one who has made space for Mahalakshmi's complete blessing.*

India did not fail to invent the financial systems of the modern world. India understood, long before those systems existed, exactly what they would do to human beings.

**And chose Mahalakshmi instead.**

*In the next chapter, we will go back further — to the oldest story India ever told about what happens when the most powerful beings in the universe decide to churn the cosmic ocean in search of the ultimate treasure. We will discover that this ancient story is not mythology. It is the most accurate warning about the nature of invention — and its consequences — that any civilization has ever produced.*

CHAPTER FOUR

# Samudra Manthan

*The Original Warning*

*The ocean already contained everything. The churning was desire.*

*— A reading of the Vishnu Purana (ancient Indian text)*

## The Story the World Calls Mythology

There is a story that India has been telling for thousands of years. It is found in the Vishnu Purana (ancient encyclopedic text), the Bhagavata Purana, and the Mahabharata (ancient epic). It has been painted on temple walls, sculpted in stone, depicted in art across Southeast Asia, taught to children before they could read, and encoded into festivals celebrated by hundreds of millions of people every year.

The world calls it mythology.

I want to suggest something different. I want to suggest that this story is one of the most sophisticated warnings about the nature of human desire and the consequences of invention that any civilization has ever produced. That it was not composed as entertainment or religious instruction alone, but as a precise, carefully coded transmission of knowledge — designed to survive across thousands of years, in a form that every person in every generation could receive.

**The story is the Samudra Manthan (churning of the cosmic ocean).**

## How It Began — With Ego

The story does not begin with demons or battles or cosmic drama. It begins with something very small. Very human.

**It begins with an insult.**

Indra (king of the gods) was riding his divine elephant Airavata when he encountered the sage Durvasa. The sage offered him a sacred garland — a gift of great auspiciousness. Indra accepted it and placed it on his elephant's trunk. The elephant, irritated by the garland's strong fragrance, threw it to the ground.

Sage Durvasa was enraged. He cursed Indra: all the gods would lose their strength, their power, their prosperity. And so it was. The gods weakened. The Asuras (demons) grew powerful and took control of the three worlds.

Notice how the story begins. Not with an external enemy. Not with a natural disaster. With Ahankar (ego born from knowledge) — Indra's casual disregard, his assumption that his position made him exempt from care and gratitude — set everything in motion.

*All suffering begins with ego. Every great churning in human history begins with someone who forgot to be humble.*

## The Decision to Churn

The gods, weakened and desperate, went to Lord Vishnu (the preserver) for help. His advice: churn the Kshira Sagara (ocean of milk). Cast all the herbs and medicines and sacred substances into it. Churn it with the great mountain Mandara as the rod, using the serpent king Vasuki as the rope. From this churning, the Amrita (nectar of immortality) will emerge.

But there was a problem. The task was too large for the gods alone. They needed to make a deal with their enemies — the Asuras — and promise to share whatever emerged from the churning.

The Devas (gods) and Asuras agreed to work together. The gods held the tail of the serpent. The demons held the head. The mountain spun. The ocean began to churn.

**This is the first lesson encoded in the story. Great inventions are never the work of one side alone. Every breakthrough in human history has required collaboration between forces that fundamentally do not trust each other.**

Modern scientific research happens this way too. Universities and corporations. Public funding and private gain. Pure inquiry and commercial application. Each side pulls in opposite directions, but the churning happens because of that very tension. Without both sides pulling, the mountain does not turn. Without the pull of opposites, nothing emerges from the deep.

## The First Thing That Emerged Was Poison

Here is where the story gets terrifying — and prophetic.

As the churning began and the ocean was disturbed at its deepest level, the first thing that emerged was not nectar. Not jewels. Not treasures.

**It was Halahala (deadly poison). A poison so toxic it could destroy the universe.**

Both the gods and the demons were paralyzed. They had not expected this. They had been promised treasures. Instead the cosmos was about to be consumed by a poison they had themselves released.

In their desperation, they ran to Lord Shiva (the lord of transformation). And Shiva — without hesitation — drank the Halahala (poison) himself. He held it in his throat, neither swallowing it (which would have destroyed even him) nor releasing it (which would have destroyed everything else). His throat turned blue from the poison — which is why he is called Neelakantha (the blue-throated one).

Read this carefully. This is one of the deepest insights in any human tradition about the nature of invention itself:

*Every churning of any depth releases poison first. The poison is not a malfunction of the process. It is built into the process itself.*

Look at the modern record. Every major invention humanity has made has produced its Halahala before producing its Amrita (nectar). The industrial revolution produced extraordinary wealth — and child labor, environmental destruction, two world wars. Nuclear technology produced clean energy potential — and Hiroshima. The internet connected all of humanity — and created mass surveillance, misinformation, addiction. Social media gave everyone a voice — and weaponized loneliness and political division.

Every churning releases its Halahala first. The poison is not a malfunction. It is built into the nature of the process. India knew this. And encoded the warning in a story so vivid, so memorable, so deeply embedded in the culture that even a child who cannot read has heard it before they start school.

**Who is your Shiva? Who, in this moment, is willing to hold the poison your inventions are releasing — to keep it from destroying the world while the rest of the churning continues?**

## The Fourteen Treasures

After the Halahala was held by Shiva, the churning continued. One by one, the treasures of the cosmic ocean began to emerge. Fourteen Ratnas (precious treasures) — each one a piece of the universe's hidden wealth, brought to the surface by the combined effort of Devas and Asuras working together.

Kamadhenu (wish-granting cow) — given to the sages to provide milk and ghee (clarified butter) for sacred yajnas (fire rituals). The first treasure was for service and nourishment, not accumulation.

Ucchaihshravas (divine white horse) — taken by Bali (demon king). Raw power without wisdom tends toward the demonic.

Airavata (four-tusked white elephant) — given to Indra. Connects earth to sky, sustains the water cycle. Balance between the terrestrial and cosmic.

Kaustubha Mani (most precious jewel) — placed on Vishnu's chest, over his heart. The greatest treasure rests not in vaults but at the heart of the preserver.

Kalpavriksha (wish-fulfilling tree) — gives whatever is asked of it, but only to those pure enough to ask for the right things.

Lakshmi (goddess of prosperity) — emerged and chose Vishnu as her consort. Prosperity aligns itself with those who preserve balance, not those who constantly create or destroy.

Varuni (goddess of intoxication) — taken by the Asuras. Intoxication, pleasure, the blurring of boundaries — claimed by the demonic principle. A precise observation about addiction and excess.

Chandra (the Moon) — claimed by Shiva and placed in his matted hair. The mastery of cycles requires the same stillness that the mastery of poison requires.

Sharanga (divine bow) — given to Vishnu. Power in service of preservation rather than domination.

Dhanvantari (divine physician) — finally, holding the pot of Amrita (immortal nectar). The last to arrive. The most precious. The physician who removes Raag (desire) itself — who was saluted in the very first verse of the Ashtanga Hridayam (Heart of the Eight Branches of Medicine) that opened Chapter One of this book.

## What the Ocean Already Contained

Here is the question that has stayed with me since I first truly understood this story:

**Where did the fourteen Ratnas (treasures) come from?**

They were not created by the churning. They were not manufactured by the Devas or the Asuras. They emerged from the ocean. They were already there — in the depths, waiting. The wish-granting cow, the divine physician, the goddess of prosperity, the Amrita. All of it. Already present in the cosmic ocean, long before anyone decided to churn.

*The churning did not create the treasures. The churning revealed them.*

This is the deepest teaching of Samudra Manthan (cosmic churning). And it connects directly to the central argument of this book.

*Everything already exists. The task is not to create — but to churn wisely, with the right intention, at the right depth, with someone willing to hold the poison that inevitably rises first.*

When Brahma (the creator) remembered Ayurveda (ancient Indian health science), he did not invent it. It was already in the cosmic ocean. When the greatest discoveries in science arrive fully formed in a single moment of insight — the apple falling, the bath overflowing, the dream that solved a problem the waking mind could not solve — are the people receiving them creating something new? Or are they churning their own inner ocean, and something that was already there is rising to the surface?

**India believed it was always the latter. Knowledge is not invented. It is churned. Remembered. Received.**

## The Gita — Samudra Manthan of the Mind

The Bhagavad Gita (the divine song — India's most celebrated philosophical text) is the dialogue between Krishna (the divine guide) and Arjuna (the great warrior) on the battlefield of Kurukshetra (the field of righteous action). Arjuna is paralyzed. He cannot fight. Not because he lacks the skill — but because Raag (desire/attachment) has collapsed him. His Moha (attachment) to his family and his Abhinivesha (fear of death) have made the greatest warrior of his age helpless.

When you look at this dialogue through the lens of Samudra Manthan, something extraordinary becomes visible:

Arjuna's questions — are the churning beginning. The mountain of crisis plunging into the ocean of a human mind.

The first rising of anguish and confusion — is the Halahala. The difficult, uncomfortable, paralyzing questions that surface before any clarity arrives.

Krishna as Neelakantha (blue-throated one) — holds the poison of Arjuna's despair. He neither dismisses it nor is consumed by it. He stays in perfect stillness while guiding the churning deeper.

The Amrita that emerges — is the understanding of Dharma (right action), of Karma (the law of action and consequence), of the soul's true nature. Not handed to Arjuna from outside — but churned from within his own consciousness through the eighteen chapters of dialogue.

**The Gita (divine song) is Samudra Manthan applied to the human mind. And the wisdom it produces — like all Ratnas from the cosmic ocean — was always already there within Arjuna. Krishna did not give him new knowledge. He helped him churn it to the surface.**

## Marma — Activating the Churning Within the Body

The Samudra Manthan is not only a cosmic event and not only an inner dialogue of the mind. It also happens at the level of the physical body.

Ayurveda describes 107 Marma points (vital energy points) across the human body — junctions where Prana (life force) is concentrated and most accessible. These points are where the body's internal electrical system — its Nadi (energy channel) network — is most sensitive and most responsive.

When a Marma point is activated through Marma Chikitsa (Marma therapy) — through precise pressure, through mantra (sacred sound), through specific yogic practice — something extraordinary happens. The body's own healing intelligence is churned to the surface. The Dhanvantari within — who is always already there, in the depths of your own biological system — begins to emerge.

Not just Dhanvantari. All the Ratnas. Health rises. Clarity rises. The natural intelligence of the body — which knows exactly what it needs, how to heal itself, how to restore balance — all of it already present in the ocean of the human body, waiting for the right churning to bring it to the surface.

**This is why Ayurveda is a system of remembering and restoring rather than a system of introducing external agents. The medicine is already within. The Marma (vital point) is simply the churning rod.**

## The Fight for the Nectar

When Dhanvantari finally emerged with the pot of Amrita, what happened?

**The Asuras immediately snatched it and ran.**

They had worked alongside the Devas. They had held their side of the churning. They had endured the poison fumes from Vasuki. They had an agreement to share whatever emerged. And the moment the greatest treasure appeared — the agreement was forgotten. The Raag that had been temporarily suppressed in service of the shared goal instantly reasserted itself.

This too is a precise description of what happens in the modern world. Two companies collaborate on research. The moment a breakthrough is achieved — the lawsuits begin over who owns the patent. Nations cooperate on scientific projects. The moment the results are useful — the competition for control resumes. The history of modern invention is full of this pattern: collaboration for discovery, conflict over ownership.

The Asuras took the nectar but could not benefit from it — because Vishnu, in the form of Mohini (a divine figure of beauty and distraction), took it back and distributed it only to the gods.

*The deepest wisdom is never successfully stolen. It can be grasped, but not possessed. It can be taken, but not used. Because the nectar of true knowledge — Amrita — only works for those who are prepared to receive it.*

## The Physician Who Emerged from the Ocean

In Chapter One of this book, we opened with the first verse of the Ashtanga Hridayam — Vagbhata's salutation to the physician who destroyed Raag without any residue. Now we know who that physician is.

Dhanvantari. The one who emerged from the churning of the cosmic ocean. The last to arrive. The most precious.

But here is what to hold in mind as we move forward through this book. Dhanvantari carried the Amrita. And the Ashtanga Hridayam tells us that the disease Dhanvantari cures is Raag itself.

**The nectar of immortality and the medicine against Raag are the same thing.**

Remove Raag — and what remains is your own undying nature. Your own Purnam (wholeness). Your own immortality, which was never created by any invention and cannot be taken by any death.

## The Warning for Right Now

Samudra Manthan was not an event from a long-ago mythology. It is happening, right now, in the most powerful technologies of the modern age.

Artificial intelligence is a churning. Genetic engineering is a churning. Quantum computing is a churning. Nuclear fusion research is a churning. Each one is a great mountain plunging into a deep ocean. Each one is pulled by forces that fundamentally do not trust each other — corporations and universities, governments and citizens, idealists and profiteers.

And each one will release Halahala before it releases Amrita. It is already happening. Artificial intelligence is producing extraordinary capability — and is also producing surveillance, misinformation, addiction, the displacement of human meaning. Genetic engineering is producing the possibility of curing inherited diseases — and the possibility of designer humans, biological weapons, ecological catastrophes we cannot predict.

**Where is our Shiva? Who is willing to hold the Halahala that this generation's churning is releasing, so that the next generations have any chance of receiving the Amrita?**

India's answer would be the same it has always been. Not a single hero. Not a single regulator. A civilizational commitment to stillness in the face of churning. The deliberate cultivation of leaders — within institutions, within governments, within communities — who can sit with the discomfort of the rising poison without rushing to release it on someone else.

The story India encoded into Samudra Manthan thousands of years ago is the most accurate warning ever produced about the moment humanity is now living through.

## What Comes Next

The first treasure to emerge from the cosmic ocean was Kamadhenu (the wish-granting cow). The cow came before Lakshmi (prosperity). Before the moon. Before the divine physician. Before the nectar of immortality itself.

This is not a coincidence. India placed the cow at the foundation of everything that followed — at the foundation of nourishment, sacred ritual, and the daily life of the entire civilization. To the modern world, this looks like superstition. To Indian civilization, it was the encoded recognition that one specific animal — a complete, self-regulating, generative system — could be studied as a model for the universe itself.

In the next chapter, we will go to the cow. And we will discover, in the simplest of farm animals, one of the most sophisticated philosophical observations any civilization has ever made about the nature of complete systems.

CHAPTER FIVE

# The Cow and The Universe

*Nature's Most Complete System*

गावो विश्वस्य मातरः

*The cow is the mother of the world.*

*— Ancient Sanskrit (ancient Indian language) saying*

## The Animal the World Misread

No aspect of Indian civilization has been more mocked, more misunderstood, and more confidently dismissed by the outside world than India's relationship with the cow.

For two hundred years, the Western world used the Indian veneration of cows as shorthand for backwardness. As proof that a civilization which worshipped an animal could not be taken seriously as a source of wisdom or knowledge.

In this chapter, I want to show you what was actually being said. Not in the language of religion — though the religious dimension is real and profound — but in the language of systems, of ecology, of the deepest science of how life sustains itself.

*The most sophisticated technology is not the one that produces the most. It is the one that wastes the least.*

By that measure, the cow is the most advanced technology India ever encountered. And India spent five thousand years studying it, protecting it, building an entire civilization around what it understood.

## The Universe Within the Cow

The Bhavishya Purana (ancient prophetic text) and other ancient texts describe something that, on first reading, sounds purely mythological: every deity in the Hindu pantheon resides within the body of the cow.

But read this more carefully. In the specific, precise mapping that the texts provide — which deity resides where in the cow's body — something extraordinary emerges.

In the mouth of the cow — all four Vedas (ancient sacred texts of knowledge). What enters through the cow's mouth is transformed by a living intelligence into something nourishing for all.

In the eyes — the Sun and the Moon. The cow sees with the light of both day and night. Its gaze encompasses the full cycle.

In the ears — the Ashwini Kumars (divine physicians/healers). The cow receives wisdom and health.

In the throat — Lord Vishnu (the preserver). The point of transformation — where what enters becomes what nourishes.

In the hooves — the mountains. What touches the earth carries the weight and stability of the Himalayas.

In the pores of its skin — the Rishis (sages/seers) and their wisdom. Every breath the cow takes carries the understanding of those who comprehended the universe.

In the dung and urine — Goddess Lakshmi (goddess of prosperity) herself. The excretion of the cow is where prosperity lives.

This is not random mythological decoration. This is a precise, systematic description of why the cow matters — encoded in the language of deity rather than the language of chemistry or ecology, because that was the language that could survive across thousands of years without distortion.

## Dattatreya and His Cow

There is one figure in Indian tradition whose relationship with the cow deserves special attention here.

Dattatreya — described in the Srimad Bhagavatam (ancient encyclopedic wisdom text) as the simultaneous incarnation of Brahma (the creator),

Vishnu, and Mahesh/Shiva (the transformer) in a single human form — is always depicted with a cow standing behind him and four dogs at his feet.

The cow with Dattatreya represents Kamdhenu (the wish-fulfilling cow) — Mother Earth herself. The one who gives everything without condition. The four dogs represent the four Vedas — always accompanying the one who has truly understood.

But the deeper meaning is this: Dattatreya, who learned from twenty-four Gurus (teachers) in the natural world — from the earth, the water, the air, the fire, the sky, from animals and human beings — always kept the cow as his companion. Because the cow is the living embodiment of the greatest teaching: give everything, retain nothing, waste nothing, harm nothing.

**The cow was not incidental to Dattatreya's wisdom. The cow was the proof of it.**

## Nature's Most Perfect Cycle

In Chapter Six of this book, we will explore the fundamental law of nature that governs all living systems: the consume-excrete cycle. Everything that lives consumes something and excretes something. And in a healthy, balanced system, what is excreted becomes the input for something else. Nothing is wasted. Everything feeds the next cycle.

**No animal on earth demonstrates this principle more completely than the cow.**

The cow consumes grass — the most abundant, self-renewing resource on the planet. It requires no rare materials, no complex supply chains, no industrial processes. Just soil, sunlight, and water.

From this simple input, the cow produces:

Milk — the most complete food known to human civilization. The only food that a newborn human being can survive on alone. From milk comes Dahi (curd/yogurt), Makhan (butter), Ghee (clarified butter) — each one with its own medicinal and nutritional properties described in extraordinary detail in Ayurvedic (ancient Indian health science) texts.

Dung (Gobar) — one of the most effective natural fertilizers known to traditional agriculture. Cow dung contains rich communities of beneficial microorganisms that help regenerate depleted soil, suppress harmful pathogens, and contribute to crop yields without synthetic chemicals. Some studies have explored its unique microbiological properties, though much of this research remains preliminary.

And at the end of its life — even then — the cow gives. Its hide. Its bones. Nothing is discarded. Nothing goes to waste. The entire body, in life and in death, is in service to the ecosystem around it.

Now compare this to the modern factory. It takes rare materials from the earth. It consumes vast amounts of energy. And what does it excrete? Waste heat. Carbon dioxide. Toxic byproducts. Electronic waste. Pollution in the air, the water, the soil. Its excretion does not nourish the next cycle. It poisons it.

**India looked at the cow and saw the model. The template. The living proof of how a system should operate if it is to be sustainable across thousands of years.**

## The Cow at the Intersection

There is a deeper layer to all of this — a way of seeing that connects the cow to the woman to the earth in one continuous principle. India recognized that all three operate on the same essential law of life: the law of nourishment without condition.

The cow gives milk that nourishes other species. The earth gives soil that nourishes every living thing that grows from it. The woman gives the womb in which a new human being develops. All three are nourishing principles. All three give without keeping account. All three have their limits — a capacity beyond which they cannot be pushed without damage. And all three, when treated with respect, can sustain life across extraordinary spans of time.

Kamadhenu (wish-fulfilling cow) is a mother. The earth is a mother. The Indian woman who manages the household, the kitchen, the children, and the seasonal practices is, in her own quiet way, a mother to the entire system she sustains. The fuller exploration of what it means that we live inside a living womb — that we are participants in a system rather than its masters — belongs to a later chapter. Here it is enough to recognize what India saw when it looked at the cow.

**Not a resource. A participant. Not a commodity. A member of the family that sustains all life.**

## Kamadhenu — The Wish-Fulfilling Cow

In the previous chapter, we met Kamadhenu (the wish-fulfilling cow) — the divine cow who emerged from the Samudra Manthan (cosmic churning) and was given to the Rishis (sages) to provide milk and Ghee for sacred Yajnas (fire rituals).

Kamadhenu fulfills desires. But she only stays with those who have released the desire to possess her. The moment a king tries to take her by force — as happens multiple times in the stories — disaster follows. She cannot be owned. She can only be served.

*The greatest abundance comes not to those who grasp for it — but to those who serve it. This is the law of Kamadhenu. This is the law of Dravya (flowing material wealth).*

## What Happened When We Forgot

The modern world looked at the cow and asked one question only: how much milk can it produce?

This single question — driven by Lobha (greed) — produced the modern dairy industry. Cross-bred cows, engineered for extraordinary milk production. Animals that produce many times more milk than they would naturally. But whose milk has different fat and protein compositions than the original Indian cow. Whose Gobar (dung) has lost many of the properties — the microorganisms, the qualities — that made the original so extraordinarily valuable. Whose connection to the natural cycle has been entirely severed.

**In optimizing for one output, the modern world destroyed the entire system.**

This is the pattern that Samudra Manthan warned about. This is Raag (desire/attachment) operating on an entire industry. The Lobha to produce more applied to the most perfect natural system ever observed, with predictably destructive results.

## The Cow and the Soul's Journey

There is one more dimension of the cow's significance in Indian tradition that must be addressed here.

Indian scriptures describe the cow as being at a unique point in the soul's evolutionary journey. According to this understanding, a soul passes through 8.4 million forms of life before reaching human form. The cow is one of the final forms before human birth — a form of relative peace, rest, and natural harmony. A kind of cosmic preparation before the demanding journey of human consciousness.

And in the rituals surrounding death — the Antyesti (last rites), the thirteen days of mourning, the sacred practices that accompany the departure of a soul — the cow plays a profound role. The ancient Indian texts describe that the cow's soul accompanies the departed soul on the journey to Vaikuntha (the divine realm/heaven). This journey, according to the Vedic (ancient knowledge) tradition, takes approximately one year to complete. The cow serves as a guide and companion through this transitional passage.

This is why the gift of a cow in Indian tradition — Go-daan (donation of a cow) — is considered among the most sacred acts one can perform for a departed soul. It is not symbolic charity. It is practical provision for a journey that the tradition understands as real.

Whether one holds these beliefs literally or reads them as a sophisticated symbolic system — the underlying message is the same: the cow is not simply an economic resource. It is a participant in the full spectrum of conscious life.

**A civilization that understood the cow this way did not need to invent factory farming. It had already found something more valuable: a living, breathing teacher of how a complete system operates in harmony with everything around it.**

## Vasudhaiva Kutumbakam — The World Is One Family

Vasudhaiva Kutumbakam (the world is one family). This phrase, from the ancient Maha Upanishad (philosophical text), is one of the most widely quoted expressions of Indian philosophy.

**But it is not merely a sentiment of universal brotherhood. It is an ecological statement of extraordinary precision.**

Vasudhaiva Kutumbakam (world is one family) says: the entire world — every human being, every animal, every plant, every microorganism (smallest organism), every river and mountain and ocean — is one family. One system. One womb.

This is the principle that makes everything in this chapter cohere. It is why the cow is venerated. It is why the earth is a mother. It is why the woman is honoured as a living expression of Shakti (feminine creative energy). It is why every living being has a place in the system, and why no living being can be reduced to a resource without the system as a whole beginning to fail.

The cow is not worshipped despite being an animal. The cow is honoured because India recognized it as a family member. A participant in the shared project of maintaining the earth's living balance. A being whose contribution to the system — through its milk, its Gobar, its presence in the field, its role in the ritual cycle — was understood to be as essential as the contribution of the farmer, the Vaidya (healer), or the Guru (teacher).

And not only the cow. Every species in the family has its assignment. Every river, every mountain, every microorganism in the soil. The Indian farmer who thanks the earth before planting, the family that addresses the river as Ganga Mata (Mother Ganges), the village that protects a particular tree as sacred — all of these are practical expressions of the same understanding. The family is real. The membership is universal. The duties of care extend to every member.

**Gavo Vishwasya Matara (the cow is the mother of the world). Not a resource. Not a commodity. Not an asset to be optimized. A mother. A nourisher. A participant in the great, ancient, endlessly self-sustaining cycle of life.**

## What India Knew

### The world spent two centuries laughing at India for worshipping a cow.

In those same two centuries, it created factory farming, chemical agriculture, antibiotic-resistant bacteria, depleted soils, contaminated waterways, and a global food system that feeds billions of people while simultaneously destroying the ecological foundation on which all food production depends.

And now — right now, in research laboratories and agricultural institutes around the world — scientists are studying Gobar (cow dung) for its microbiological properties, traditional Indian farming methods for their soil regeneration effects, and the ancient concept of Panchagavya (the five products of the cow) for their potential applications in medicine and agriculture.

**The world is, slowly and reluctantly, catching up to what India always knew.**

*India did not fail to invent factory farming. India understood the cow too well to ever reduce it to a factory.*

*In the next chapter, we will move from the visible to the invisible. From the cow — whose gifts you can touch and taste and smell — to something that runs through every living being on this planet but that no human eye has ever seen. We will talk about electricity. Not the kind that comes from a power station. The kind that is already inside you.*

CHAPTER SIX

# The Electricity Within You

*The Invention India Never Needed*

*The most powerful electrical system you will ever encounter is not in any laboratory. It is in your body.*

*— A reading of the ancient Vedic understanding through the lens of modern bioelectrical science*

## The Box in Your Kitchen

### You have pressed that button a thousand times.

You have stood in front of the microwave and watched your food heat up — from nothing. No flame. No visible source of heat. Just a box humming quietly while something invisible passes through your food and makes it hot.

What is happening is this: invisible waves of electromagnetic radiation (energy waves) are passing through your food, vibrating the water molecules within it at billions of times per second, generating heat through that friction alone.

You cannot see the waves. You cannot feel them. They are completely invisible to every human sense. And yet they are real. Measurable. Powerful.

Now — here is the question India was always asking, that almost nobody in the modern world has stopped to ask:

*If invisible waves can heat your food — what are the invisible waves passing through YOU doing to your body, your mind, and your consciousness right now?*

This chapter is about electricity. But not the electricity that comes from a power station. The electricity that was already there — long before any of those things existed. The electricity that runs inside you. That India spent five thousand years learning to understand, to master, and to use.

## The Law That Everything Follows

In the previous chapter, we explored the consume-excrete cycle: everything that lives consumes something and excretes something. In a healthy, balanced system, the excretion of one organism becomes the input for another. Nothing is wasted. Everything feeds the next cycle.

A human being consumes food, water, air. Excretes waste, carbon dioxide, heat. A cow consumes grass. Excretes milk, Gobar (dung), Gomutra (urine) — each feeding the next cycle. A tree consumes sunlight, water, carbon dioxide. Excretes oxygen, fruit, fallen leaves that become soil. Complete.

Now look at the electrical appliances that the modern world invented. A car consumes petrol. Excretes carbon dioxide, carbon monoxide, nitrogen oxides, heat — none of which nourish anything. A mobile phone consumes electricity. Excretes electromagnetic radiation continuously, heat, and eventually becomes electronic waste containing toxins that leach into soil and groundwater for decades.

**The pattern is consistent: every electrical appliance that humanity has invented breaks the fundamental law that every natural system follows. Its excretion does not nourish. It accumulates as poison.**

## The Leaf and the Sun

Here is an observation that seems simple but contains something profound.

A tree leaf absorbs full sunlight for the entire day. Every day. In summer and in winter, in heat and in rain. The same sun that increasingly damages modern human skin — that causes burns, premature aging, and skin cancer when met by an unprepared body — falls on that leaf continuously, and the leaf flourishes.

The leaf is more delicate than human skin in many ways. A single touch can damage it. A single insect can tear it. And yet, day after day, the most powerful energy source in our solar system falls on it, and the leaf does what it was designed to do.

The leaf is not invulnerable. It has its own protective mechanisms — pigments and biochemical defences against intense light. It is not that the sun has no effect on the leaf. It is that the relationship between the leaf and the sun is, fundamentally, a relationship of partnership rather than antagonism.

**The leaf does not fight the sun. It uses it. It converts solar energy directly into life through Photosynthesis (the process of converting sunlight to food). The sun is not the leaf's enemy. The sun is the leaf's food.**

The human body was also designed to work with the sun. Not against it. Our bodies produce Vitamin D (essential nutrient) through sun exposure. Sunlight regulates the Circadian rhythm (internal biological clock), governs serotonin production, and supports immune function. We evolved under the sun for millions of years. The relationship is, by default, partnership rather than antagonism.

The ancient Indian farmer worked in the fields under the same sun for hours every day — and met it differently than the modern body does. Not because his skin was invulnerable; sun wears on every human body. But his exposure was gradual, daily, and year-round rather than sudden and rare, and his relationship with the sun was intact. His diet — rich in turmeric (a natural anti-inflammatory spice), ghee (clarified butter), antioxidant-rich spices and seasonal foods — protected his skin from within. His mind was calm. His body was in its natural rhythm. He was, like the leaf, aligned with the sun rather than at war with it.

Modern human beings often avoid the sun completely, then suddenly expose pale, nutritionally depleted, chronically stressed bodies to intense ultraviolet radiation — and wonder why the damage occurs. It is not the sun that changed. It is the human's relationship with the sun that broke.

None of this is an argument against protecting the skin — protection matters, and medical guidance on sun exposure exists for good reasons.

**The leaf never invented sunscreen. Not because the sun cannot affect it, but because it never broke its relationship with the sun in the first place.**

India's ancient practices — seasonal diet, daily sun exposure during the correct hours, the morning Surya Namaskar (salutation to the sun), the agricultural life that kept human beings connected to the full cycle of daylight — maintained this relationship. Sunscreen protects, and the protection is real. But it is protection for a body whose relationship with the sun has already been broken — a remedy applied downstream of a rupture that a different way of living did not create in the same way.

## The Mop and the Sun — Ancient Antiseptic Science

Here is another observation connected to the same understanding.

Take a wet mop or cloth that has been used to clean a floor. If you leave it in a dark, damp corner, it becomes a breeding ground for bacteria within hours. The smell alone tells you what has happened. Left long enough, the microbes that colonize it can spread across the floor it was meant to clean, across hands that touch it, through an entire household.

Now take the same mop. Instead of washing it with soap and chemical disinfectant — hang it in direct sunlight for a few hours.

The ultraviolet radiation in direct sunlight damages the DNA of bacteria, viruses, and many other microorganisms. It is one of the most effective natural disinfectants on earth. No chemicals. No residue. No cost. The mop can be used again safely.

Every Indian grandmother knew this. The practice of drying bedding, clothing, and cleaning materials in direct sunlight was not a matter of custom or convenience. It was applied antiseptic science — the practical use of sunlight as a purifying agent, passed from generation to generation through daily practice rather than through any textbook.

Modern hospitals now use UV light to sterilize operating rooms. Modern water treatment systems use UV radiation to purify drinking water. Modern food processing uses UV light to eliminate pathogens from surfaces.

**India's grandmothers were using the same fundamental principle on their mops and their mattresses and their yards for thousands of years before the first hospital UV lamp was ever switched on.**

The knowledge was not absent. It was encoded in practice. And that practice is grounded in real biology. Sunlight inactivates microbes. Sunlight, when the human body is properly prepared to receive it, supports health. The grandmother and the modern microbiologist are looking at the same fact through different windows.

## The Invisible Storm

**The intensity of electromagnetic radiation in the human environment is now at levels never before experienced in the history of life on this planet.**

Before the industrial revolution, the electromagnetic environment of the earth consisted of natural sources: sunlight, lightning, the earth's own magnetic field. The human body evolved over millions of years within this environment. It knows how to work with these natural electromagnetic fields.

In the space of approximately one hundred and fifty years — a blink in evolutionary time — humanity has added to this natural background an almost incomprehensible storm of artificial electromagnetic radiation. Radio waves. Microwave radiation from phones, routers, cell towers. Electrical fields from power lines and household wiring. The combined output of billions of phones, computers, televisions, and wireless devices operating simultaneously around the clock.

The full health implications of this exposure are still being studied. Some research has explored how sustained exposure to artificial electromagnetic fields may interact with the body's own electrical signals — its nerve function, its cardiac rhythm, its sleep patterns. The findings are mixed and the science is evolving. But what is not in dispute is the scale and the speed of the change. The electromagnetic environment of the human body in the modern world is fundamentally different from the one in which the human body evolved. Whether and to what degree this matters is a question that will take many more years to answer fully.

The microwave heats your food by vibrating its water molecules at a specific frequency under specific conditions. Microwave ovens are designed to contain that radiation within the appliance, and within those design parameters they are considered safe. But the broader question — what an environment saturated with overlapping artificial electromagnetic fields means for a body that was never tested under these conditions — remains an open one.

**India, thousands of years ago, was already asking the question that the modern world is only now being forced to take seriously.**

## Vayu — The Original Communication Medium

There is something that every wireless technology in the modern world has in common. Radio. Television. Mobile phones. WiFi. Bluetooth. Satellite communication. All of them — without exception — use the same medium to carry their signals.

**Air.**

Or more precisely — the electromagnetic field (energy field) that permeates air and all of space. Every signal your phone receives has travelled through the air to reach you. Every broadcast, every call, every data packet — all of it carried by the same medium that India always called Vayu (air/wind — the second of the five Panchabhuta/five elements).

India understood Vayu (air) as a carrier — not merely of oxygen, but of Prana (life force), of sound, of vibration, of the deepest signals of the universe. The Vedic (ancient knowledge) texts describe Vayu as the primary medium through which consciousness communicates across space.

And Hanuman (son of Vayu/Pawan Putra — son of the wind) — the master of Vayu — is described in the Ramayana (ancient epic of Rama) as being able to travel anywhere instantly, to know what is happening across vast distances, to carry messages and information with perfect fidelity across the ocean itself. He is the original wireless signal carrier.

The modern telecommunications engineer and the ancient Vedic sage were both describing the same phenomenon: that information crosses distance through a pervading medium. The engineer calls it the electromagnetic field — needing no air at all, crossing even the vacuum between worlds. And the tradition, remarkably, had already drawn the same line: the sound we hear rides on Vayu, but the subtle substratum of all vibration is Akasha — space itself, the first element — which the classical schools describe as one, eternal, all-pervading, and imperceptible. The engineer's field and the sage's Akasha are pointing at kindred territory. One described it in the language of electromagnetic frequency. The other in the language of Prana and divine power.

**The modern world built towers and transmitters to use this medium. India encoded the understanding of this medium in its greatest devotional figure — Hanuman — whose name is chanted by hundreds of millions of people, whose mantra (sacred sound) travels through Vayu every day.**

## India's Electrical Science

**The human body is not simply affected by electricity. The human body IS an electrical system. One of the most sophisticated ever to exist.**

Every thought you have is an electrical signal passing between neurons. Your heart beats because of a precise electrical pulse. Your muscles contract because electrical signals travel along nerve fibres at speeds of up to 120 metres per second. The ECG (electrocardiogram — heart electrical reading) machine does not generate electricity in your heart. It reads the electricity that is already there.

Modern science has taken approximately two hundred years to develop the instruments needed to read these signals.

**India spent five thousand years learning to work directly with them.**

The concept of Prana (life force/vital energy) describes the fundamental bioelectric current that animates all living systems. Five primary forms of Prana govern the five major functions of the body: the intake of energy, its distribution, its metabolic processing, its excretion, and the upward movement of consciousness.

The Nadi system (energy channel network) — described in yogic texts as a network of seventy-two thousand channels through which Prana flows — maps the body's pathways with a precision that modern neuroscience continues to investigate. The three primary Nadis — Ida (cooling/lunar channel), Pingala (heating/solar channel), and Sushumna (central channel) — correspond closely, in their functional descriptions, to what modern science understands as the parasympathetic (calming) nervous system, the sympathetic (activating) nervous system, and the central nervous system.

The Chakra system maps regions of the body that are, in physiological terms, dense with nerve plexuses and endocrine activity — a correspondence the tradition described in its own vocabulary long before anatomy had one.

The Marma points (107 vital energy points of Ayurveda/ancient Indian health science) correspond, in many cases, to nerve junctions, vascular confluences, and bioelectrically active zones that acupuncture research and studies in bioelectrical medicine have identified as sites of heightened sensitivity.

And the practices India developed for working with this internal electrical system:

Pranayama (breath regulation) — research has shown that controlled breathing patterns can modulate the autonomic nervous system (self-regulating nervous system), alter heart rate variability, change brain electrical activity, and reduce cortisol (stress hormone) levels.

Yoga (union/body-mind practice) — the precise positioning of the body in specific configurations that alter the distribution of bioelectric current through the system. Every asana (posture) is, from a bioelectrical perspective, a different circuit configuration.

Dhyana (meditation/focused attention) — EEG (brain electrical reading) studies of experienced meditators have shown distinctive patterns of coherent electrical activity — particularly in the gamma frequency range — associated with heightened awareness and integrated cognition. This is one of the most well-replicated areas of contemplative neuroscience.

Mantra (sacred sound vibration) — sound vibration used deliberately to affect the body's bioelectric state. Specific sound frequencies create specific vibrational patterns in the body's tissues and fluids. Vedic sages mapped which sounds produce which effects, and encoded this knowledge in the precise syllables, rhythms, and tones of sacred chanting.

## The Question India Was Always Asking

Imagine two civilizations standing side by side, both looking at the same question: how do we harness the power of electricity to serve human life?

The modern world looked outward. It built generators and turbines and power stations and wires. It created an external electrical system of extraordinary scale and power. But the external electrical system has a cost. It saturates the electromagnetic environment with frequencies the body was never designed to navigate.

India looked inward. It asked a different question:

*What if the most powerful electrical system we will ever need is already inside us? What if the work is not to build an external system — but to understand, to clear, and to master the internal one?*

India also understood that it was not possible to build an external electrical grid without disrupting the interdependence of all living systems. Birds navigate by the earth's electromagnetic field. Bees find flowers through electrical signals. Fish orient themselves through bioelectric sensitivity. Trees communicate through underground fungal networks that carry electrical impulses. Every living system on earth is embedded in a web of natural electromagnetic communication, and any artificial external system has the potential to disrupt that web in ways we are still learning to understand.

Whether or not anyone ever weighed the decision in these terms, a civilization that saw every living system as embedded in one web had no framework in which blanketing that web with artificial fields could look like progress.

## Your Body Has All the Answers

There is a piece of wisdom that has been passed through Indian households for generations. It is not from a scripture or a philosophical text. It is something older than any text — something that the body itself teaches, if you are still enough to listen:

*Aapki body hi aapke sab prashno ka jawab degi. Aap shant raho, ghar me jhadu aur pocha karo.*

*Your body will answer all your questions. Stay calm. Sweep and mop your home.*

This may sound simple to the point of being dismissible. But look at what it is actually saying.

The body has all the answers. Not the internet. Not the library. Not the expert. Not the algorithm. The body — the electrical system running inside you right now, which has access to extraordinary information through the consciousness that animates it.

But the body can only speak when the noise is quiet enough to hear it. When Raag (desire/attachment) is not drowning out the signal with its endless wanting and fearing and comparing. When the electromagnetic storm outside is not interfering with the delicate electrical conversation happening inside.

Stay calm. This is Pranayama, Dhyana (meditation), stillness. The deliberate creation of the internal conditions under which the body's electrical system can operate at its full capacity.

Sweep and mop your home. The physical act of jhadu and pocha — simple, repetitive, grounding, done close to the earth — brings the nervous system into a state of calm alertness. It is moving meditation. The body doing something it understands completely, freeing the mind to settle.

## Why India Didn't Invent the Power Grid

**A civilization that understood electricity this way did not need to invent the power grid.**

Not because it was unaware of electrical phenomena. Lightning was observed and understood. The electromagnetic nature of the cosmos — described in the concept of Akasha (space/ether — the fifth of the Panchabhuta/five elements) that permeates all space and carries all vibration — was a foundational principle of the entire philosophical system.

But India looked at the question of electricity and asked: before we build an external electrical system, have we mastered the internal one? Have we understood the Prana? Have we mapped the Nadis (energy channels)? Have we learned to work with the Chakras (energy centers) and the Marma (vital points) and the bioelectric fields that govern health, consciousness, and the quality of human life?

The modern world went outward first. And now, as that external system fills the electromagnetic environment with frequencies the body was never designed to navigate, the modern world is beginning to ask the question India never stopped asking:

*What is this electricity doing to us?*

**India already knew. And chose to master the electricity within before building the electricity without.**

## The Technology That Belongs to Everyone

There is something remarkable about India's electrical technology — the system of Yoga (mind-body union practice), Pranayama, Dhyana, and Mantra that it spent five thousand years developing.

**It is copyright free.**

Nobody owns Yoga (mind-body practice). Nobody owns Pranayama. Nobody owns Dhyana. These technologies were developed, refined, and given freely to anyone willing to practice them. There is no patent on Surya Namaskar (sun salutation). There is no proprietary algorithm in the breath.

Hundreds of millions of people around the world practice Yoga today. They feel the results: reduced stress, improved sleep, greater clarity, better physical health, a sense of connection to something larger than the ordinary noise of daily life. They are, without necessarily knowing it, doing exactly what India always said to do: working with the electricity within, rather than simply consuming the electricity without.

No other civilization in human history produced a complete, systematic, freely available technology for mastering the human body's internal electrical system and gave it, without restriction, to all of humanity.

*India did not invent the light bulb. India invented the technology for becoming your own light.*

*In the next chapter, we will look at the inventions India DID make — the ones the world cannot see, touch, or patent. The technologies that exist not in metal or silicon but in sound, geometry, and the fabric of consciousness itself. We will discover that India was not a civilization that failed to invent. It was a civilization that invented things so profound that the world is only now developing the instruments needed to measure them.*

CHAPTER SEVEN

# The Inventions You Cannot See

*Mantra, Yantra and Tantra — India's Original Technology*

## What the Archaeological Record Misses

When historians measure the inventiveness of a civilization, they look for physical evidence. Tools. Machines. Structures. Written texts that describe new processes or technologies. The wheel, the printing press, the steam engine — all of these left evidence that could be excavated, dated, and catalogued.

By this measure, India appears to have invented relatively little in the modern sense. And this appearance has been used, repeatedly, as evidence of India's intellectual or civilizational limitations.

But this measure has a fundamental flaw. It can only detect the kind of invention that leaves physical evidence. It is completely blind to another category of invention entirely — one that India pursued with extraordinary depth and sophistication for five thousand years.

The invention of technologies that work through consciousness. Through sound. Through geometry. Through the invisible structures of the universe itself.

**India's three greatest invisible inventions were Mantra (sound tool of the mind), Yantra (geometric instrument), and Tantra (universal network/method). Together, they form a complete technology of consciousness — a system for working with the invisible forces that govern all of existence.**

## Mantra — The Original Wireless Communication

The Sanskrit (ancient Indian language) word Mantra (sound tool) comes from two roots: Man, meaning mind, and Tra, meaning instrument or tool. A Mantra is literally a tool of the mind — a precisely engineered sound sequence designed to produce specific effects in the human bioelectric system and in the environment around it.

Sound is vibration. Vibration is energy. Specific vibrational frequencies interact with physical matter in measurable ways. The science of cymatics (study of visible sound patterns) demonstrates that sound waves can organize matter into precise geometric patterns. Different frequencies produce different patterns. Change the frequency, and the pattern changes.

The Vedic (ancient Indian knowledge) Rishis (sages/seers) understood this. They spent thousands of years mapping which sound frequencies produce which effects — in the human body, in the environment, in the subtle layers of consciousness. A growing body of research has examined some of these effects. Studies have suggested that regular chanting of specific Mantras (sacred sounds) can reduce cortisol (stress hormone) levels, alter brain wave patterns, and improve heart rate variability.

**India called it Mantra. Modern science calls it acoustic therapy, sound healing, and vibro-acoustic treatment. They are describing closely related phenomena. India was there first — by thousands of years.**

## How a Mantra Was Made

This raises a question that the modern mind almost never thinks to ask. Where did the Mantras come from?

The modern world produces inventions through laboratory work, peer-reviewed publication, patent filings, and commercial release. There is a clear chain of evidence from question to discovery to product. The Mantras of the Indian tradition did not arrive this way. They arrived through a completely different method — one preserved in texts like the Agni Puran (the encyclopedic text of fire and sacred ritual) and transmitted through institutions that have no real equivalent in the modern world.

The Mantra was not composed. The Mantra was received — the same way Brahma received Ayurveda in the verse we examined earlier in this book. The

Rishi (sage) entered a state of extraordinary internal stillness. The mind was emptied of its ordinary content. The breath slowed. The body was held in specific yogic configurations that altered its bioelectric state. And in that stillness, the Mantra arrived — as a specific sound-pattern that carried within it a specific energetic effect.

Once received, the Mantra was tested. It was chanted by the Rishi over years, sometimes lifetimes. Its effects on the body, on the mind, on the environment, on the people who heard it, were observed with the patience of a civilization that did not need to publish results next quarter. The Mantras that survived this process — that were verified, replicated, and refined across generations of practitioners — became part of the lineage.

Each lineage had its own Mantras. The various Maths (monastic institutions) and Akhadas (orders of ascetic practitioners) — the ancient and still-living infrastructure of Indian spiritual transmission — held specific Mantras that were not part of the general public corpus. These were the proprietary knowledge of the lineage. They were transmitted, in many cases, only orally. They were given by the Guru (teacher) to the disciple at a specific stage of preparation. They were never published. They were never written down for general circulation. They were never spoken aloud in public spaces, because the spoken sound itself carried the technology, and the technology was not meant to be loose in the world.

**Some Mantras were so carefully guarded that they were transmitted only once in a lifetime, in a single whispered transmission from teacher to student. Once whispered, the sound entered the field — and the Indian understanding was clear that what enters the field cannot be unspoken.**

This is not superstition. This is the encoded recognition of something that the modern world is only beginning to understand. Sound is not an ephemeral phenomenon. The vibration, once released into the medium of Vayu (air), travels. The Indian tradition treated voice and Mantra (sacred sound) with the care that modern physics now recognizes — that an electromagnetic signal, once broadcast, propagates outward and does not stop. The next chapter of this book will return to this in detail. For now, simply note: the Rishis (sages) treated spoken sound as if it were permanent. Because, in a sense that the modern world is rediscovering, it is.

There is one more dimension of Mantra creation that the Agni Puran (ancient text on sacred ritual) describes precisely. A Mantra is not arbitrary.

The syllables are not interchangeable. The order matters. The breath behind each syllable matters. The intention with which each is voiced matters. The Mantra operates, in the language of modern engineering, as a kind of key — a key whose precise shape opens a specific lock in the energetic structure of consciousness. The Rishi who created a Mantra was not writing a poem. The Rishi was cutting a key. And the lock the key opens has been there, in human consciousness, since before there were keys.

*Key. Lock. Vibration. Field. These are the words modern engineering uses for what the Agni Puran was describing in Sanskrit thousands of years ago.*

## Ram Se Bada Hai Ram Ka Naam

There is a statement in the devotional tradition of India that has been repeated across centuries, across languages, across every region of the subcontinent:

*Ram se bada hai Ram ka naam. The name of Ram is greater than Ram himself.*

How can a name be greater than the one it names? This question — which sounds paradoxical at first — opens into one of the deepest insights in all of Indian philosophy.

The name Ram — in Sanskrit, R-A-M — is a specific configuration of sound. A specific vibrational pattern. And when you chant it slowly and clearly — Ram Ram Ram — something extraordinary happens as the pace deepens.

The sounds begin to blur into each other. Ram Ram Ram becomes Mara Mara Mara — the word for death in Sanskrit, pointing to the dissolution of ego. And as the chanting deepens further still, even the vowels dissolve, leaving only the pure consonant vibration Rm Rm Rm — a frequency below language, below meaning, at the level of pure vibration itself.

This vibrational journey — from the external audible name, through the dissolution of ego, to the pure frequency beneath all form — is a complete spiritual technology encoded in three syllables.

Now consider AUM (the primordial sound — the first vibration of the universe). Studies have explored the acoustic and physiological effects of chanting AUM, and have observed shifts in brainwave patterns, activation of the vagus nerve (the body's main calming nerve), and changes in the parasympathetic nervous system (calming system). Much of this research is still in early stages, but it points to something that the Vedic tradition has always claimed: that sound, when used with precision, can produce real effects in the body and the mind regardless of the chanter's intellectual belief about what is happening.

**AUM and RAM, at their deepest level of pure vibration, point toward the same fundamental territory — the vibrational ground of all existence. This is why Ram se bada hai Ram ka naam (the name of Ram is greater than Ram). The physical form of Ram — however divine — exists within time. The vibrational frequency of the name transcends time.**

## The Son of Air

Hanuman is the son of Vayu (air/wind — one of the Panchabhuta/five elements) — also called Pawan Putra (son of wind/air). This is not an incidental detail of his mythology. It is the central fact of his identity. He is the child of the force that moves through all space, carries all vibration, connects all living beings to each other.

Vayu is the medium through which all communication travels. Radio waves. Television signals. Wireless internet. The voice of a mother calling her child. The Mantra of a devotee rising toward the sky. All of these travel through air. Air is not empty space. Air is the original carrier of information.

And Hanuman — born of this medium — could do things that seem impossible by ordinary physical logic. As a child, he mistook the rising Sun for a fruit and flew toward it to eat it — crossing the vast distance between earth and the solar sphere in a single bound. A being that operates in the medium of Vayu — in the medium of vibration, of consciousness — is not limited by the distances that constrain physical objects.

But there is something even more important about Hanuman's story. He forgot his own abilities as he grew older. He lived for years in a state of reduced power — not because the power was gone, but because he had forgotten it was there. Then came the moment his power was needed. And it was Jamvant (the elder, the one who remembered) who spoke the words that restored everything.

Not because Jamvant gave him new power. Because Jamvant reminded him of the power he had always had but forgotten.

**This book is a Jamvant. It is not giving you new information. It is reminding you of what you already know. The knowledge of the universe is already in you. It is simply waiting to be remembered.**

## The Eight Siddhis — States of Consciousness, Not Superpowers

Hanuman is described in the Sundarkanda (beautiful chapter — section of the Ramayana) as the master of the Ashta Siddhi (eight states of mastery) and Nava Nidhi (nine treasures). The Ashta Siddhi are:

Anima (the ability to become infinitely small) — the state of consciousness in which the ego is completely dissolved. The person who has achieved Anima (infinite smallness) has no arrogance, no claim, no insistence on being seen. They can enter any situation without disturbance because they carry no ego into it.

Mahima (the ability to become infinitely large) — the state of consciousness in which the self expands to contain all of existence. Not physical size — but the quality of presence that a truly conscious person carries. The person whose awareness is so vast that others feel held within it.

Garima (the ability to become infinitely heavy) — the state of absolute groundedness. A consciousness so stable, so rooted, that no external force — no pressure, no threat, no inducement — can move it from its center.

Laghima (the ability to become infinitely light) — the state of complete non-attachment. A consciousness so free from Raag (desire/attachment) that it is not weighed down by any outcome. It can move freely because it carries no burden of wanting.

Prapti (the ability to reach anywhere) — the state of complete access. A consciousness so clear and so present that it can understand any situation, connect with any person, perceive any truth. Not physical travel — but the quality of attention that misses nothing.

Prakamya (the ability to achieve any desire) — the state of perfect alignment between will and action. A consciousness so free from internal contradiction that what it intends, it accomplishes. Not because it forces the world — but because it moves with the natural flow of things.

Ishitva (mastery over all elements) — the state of natural leadership — not domination, but the kind of mastery that comes from understanding. A consciousness so attuned to the laws of nature that it can work with any element of existence.

Vashitva (the ability to influence all beings) — the state of genuine influence — not manipulation, not force, but the natural magnetism of a consciousness that is fully aligned with truth. People, animals, even elements respond to such a consciousness because they recognize something in it that they also carry within themselves.

These eight Siddhis (states of mastery) are not supernatural powers in the fantasy sense. They are states of consciousness — qualities of awareness — that become available when Raag is sufficiently reduced. They describe what a human being naturally IS when the noise of desire and fear has been quieted enough to access what was always already there.

**Hanuman possessed all eight — not because he was given them from outside, but because he had mastered his own consciousness completely. He carried the Ram Naam (name of Ram) in every cell of his being. The Mantra and the Siddhi (mastery) are inseparable.**

## Yantra — The Original Engineering

The Sanskrit word Yantra means that which restrains and focuses. A Yantra is a geometric diagram — a precise visual structure — that serves as a map of a specific pattern of energy in the universe.

The Sri Yantra (auspicious geometric instrument) — perhaps the most celebrated of all Yantras — is constructed from nine interlocking triangles radiating from a central point. Mathematicians and researchers who have studied the Sri Yantra over the years have noted the remarkable precision of its construction, and how difficult it is to draw correctly without an iterative process. Whether the geometry is unique in any formal mathematical sense remains debated. What is not debated is that millions of practitioners across thousands of years have used it as a focal instrument for the training of human attention.

The computer chip (microprocessor) operates on similar principles. It is a geometric arrangement of circuits designed to process information by directing electrical current along precisely engineered pathways. The circuit board is, in a sense, a Yantra for electricity.

**India's Yantra channeled not electricity but consciousness. And in doing so, it achieved results that no computer has ever replicated: the systematic training of human attention, the development of states of awareness that modern neuroscience increasingly recognizes as among the highest functioning of the human brain.**

## Tantra — The Original Network

Of the three, Tantra is the most widely misunderstood — and therefore the most important to address directly.

In popular understanding outside India, Tantra (universal network) has been almost entirely reduced to its most superficial dimension — a set of practices involving sexuality. This reduction is about as accurate as describing the internet as a system for sending emails. It identifies one small application while completely missing the underlying technology.

The Sanskrit root of Tantra means to stretch, to weave, to extend. A Tantra is a system that weaves together different dimensions of experience into a unified whole. In its broadest sense, Tantra is the science of the universal nervous system — the network that connects all consciousness, all life, all matter into one coherent field.

Every Indian ritual is, at its core, a Tantric (network-based) practice. Lighting a lamp. Offering flowers. Chanting a specific combination of syllables at a specific time of day. Performing a gesture with the hands — a Mudra (hand gesture/circuit) — that completes a specific circuit in the body's bioelectric system. Each is an input into the universal network. Each produces a response.

**Tantra is India's internet. But it operates through consciousness rather than silicon. Its reach is not limited to devices connected by cables and radio towers. It extends — according to the ancient understanding — to every conscious being in the universe.**

## Karmkand and Havan — Encoded Technology in Sanskrit

There is one form of Tantric practice that has been carried in Indian households more continuously than any other, across more centuries than any other, with less interruption than any other. It is called Karmkand (the body of ritual action).

Karmkand is the system of precise ceremonial procedures that surrounds every major life event in traditional Indian life — birth, naming, the beginning of education, marriage, the construction of a home, the laying of foundations for a business, the consecration of a temple, the rituals of death. It is a vast body of practice, transmitted across thousands of years primarily through the Brahmin lineages whose specific role in Indian society was the preservation and execution of this technology.

To the casual modern eye, Karmkand (ritual action) looks like superstition. A long sequence of unfamiliar actions, performed in unfamiliar Sanskrit, accompanied by smoke and lamps and offerings, watched by a family who in many cases cannot translate the syllables being chanted. Easy to dismiss. Easy to skip in modernizing households.

But read carefully, Karmkand is a precise composite of every technology this chapter has discussed.

It deploys Mantra — specific syllables chanted in specific orders at specific moments. It deploys Yantra — the diagrams drawn in colored powder, sandalwood paste, vermilion, and turmeric on the floor of the ritual space. It deploys Mudra — the precise hand positions of the priest and the participants that complete specific bioelectric circuits. It deploys substance — the specific herbs, fruits, grains, oils, and Ghee (clarified butter) that are placed in the fire.

And almost every major Karmkand (ritual) culminates in the same act: Havan (the offering of substances into a sacred fire). The Havan is not symbolic. The Havan is the operating mechanism of the ritual. Fire transforms matter into energy. Ghee combusted at a specific temperature releases specific aromatic compounds. The smoke from particular herbs has known antimicrobial and physiological effects when inhaled. The Sanskrit syllables chanted while the substances burn carry the sound frequencies discussed earlier in this chapter into the same vibrating air that is now warm with the energy of the fire.

And Karmkand understood something else the modern world is only beginning to measure: the power of synchrony. When millions of households perform the same act at the same hour — light the same lamp at dusk, sound the same bell, ring the same thali (metal plate) from their balconies in a moment of collective crisis — something happens that no individual act can produce. Synchronized ritual is an ancient technology of belonging: it tells every nervous system in the network, at the same instant, you are not alone.

India engineered this at civilizational scale, on a calendar, for thousands of years.

**Karmkand is a complete, multi-modal technology. Sound, geometry, substance, fire, and the human nervous system — all combined into a single sustained operation lasting hours or days, executed by a trained practitioner whose entire life has been a preparation for the execution of this specific procedure.**

The Brahmin (priestly) tradition was the engineering corps of this technology. The Vedas (ancient sacred texts) were their reference manuals. The specific procedures for each kind of Karmkand were preserved in painstaking detail in texts that were, in their original form, technical documents — not poetry. The Vedic priests were not poets reciting verse for atmosphere. They were technicians executing a procedure whose steps had been refined across thousands of years of observation.

And the Havan (fire offering) was, in this larger picture, the principal output device of the entire system. The fire was the place where the invisible became visible, where the chanted sound entered the heated air and travelled outward, where the energetic effect of the ritual was delivered into the world. India did not build factories. India built fires. And the fires were, in their own quiet way, performing the same role for the inner technology that the smokestacks of the industrial revolution would perform for the outer.

There is no modern equivalent of Karmkand. The closest analogies — a clinical surgery, the firing of a rocket, a complex computer program — give some sense of the precision involved, but none of them involves consciousness as the primary operating medium. Karmkand does. And that is why it survives. The infrastructure of the modern world will, in time, be obsolete. The infrastructure of Karmkand was designed to work in any century, on any continent, for as long as human beings carry the equipment of body and breath that is required to execute it.

## When Your Friend Appears

You have experienced this. Almost everyone has.

You are sitting with a friend. You begin talking about someone — a person who has no particular reason to contact you at this moment. And within minutes — sometimes within seconds — that person calls, or messages, or walks through the door.

In India, the traditional response is: you will live for a hundred years. But beneath this cultural response is a more precise observation: the person appeared because they were already thinking about you. Their attention moved toward you first — and that movement of attention, through what India recognized as the universal network of consciousness, reached you and triggered your conversation about them.

**The signal came before the physical arrival. This is the Tantra operating in daily life.**

## Dhanurveda — Archery Was a Veda

Before we close this chapter, there is one connection that must be made here — one that becomes important in the chapters that follow.

The complete map of India's knowledge system shows that alongside Ayurveda (ancient Indian health science) and Gandharva Veda (music, art and dance as sacred knowledge), there exists Dhanurveda (archery and warfare as sacred knowledge). Dhanurveda is an Upaveda (applied sacred knowledge) — one of the four applied branches of the Vedas (ancient sacred texts of knowledge).

This means that archery in ancient India was not a skill. It was not a sport. It was not even a martial art in the way the modern world understands martial arts.

Archery was a Veda (sacred knowledge system). A complete knowledge tradition encompassing not only the physical mechanics of bow and arrow — but the psychology of the archer, the philosophy of aim and intention, the spiritual dimensions of sending a force through space with complete attention and complete surrender.

**The Yantra of the archer was the bow. The Mantra was the internal focus that aligned consciousness with the target. The Tantra was the complete system of Dhanurveda (archery as sacred knowledge) that could be accessed not from a teacher but from within.**

**Three Technologies. One Complete System.**

Mantra. Yantra. Tantra. Three technologies that together form a complete system for working with the invisible dimensions of existence.

Now compare this to what the modern world invented:

Radio and wireless communication — uses invisible waves through air to carry information. India called this Mantra and the power of Vayu.

Computer chips and circuit boards — use geometric arrangements to direct electrical energy with precision. India called this Yantra.

The internet — a network that connects all nodes of information into one field, enabling instantaneous communication across any distance. India called this Tantra.

The modern world spent five hundred years inventing external versions of what India had already mapped internally. The external versions require power stations, server farms, undersea cables, satellite networks, and the continuous extraction of rare materials from the earth. They break down. They become obsolete. They generate the Halahala (poison) discussed in Chapter Four.

**The internal versions require nothing external. They run on the electricity within the human body. They are powered by breath. They are free. They are available to every human being alive. And they have been available, without subscription or upgrade or infrastructure requirement, for thousands of years.**

## The Patent That Was Never Filed

India never patented Yoga (mind-body practice). Never patented Pranayama (breath regulation). Never patented the Sri Yantra or the system of Tantric practice. Never charged for access to the Vedic (ancient knowledge) Mantras. Never restricted the knowledge of the Chakra (energy center) system or the Nadi (energy channel) map.

This was not oversight. This was the deepest expression of Vasudhaiva Kutumbakam (the world is one family). Knowledge that belongs to the universal consciousness cannot be owned by any individual or any civilization. It can only be received, practiced, and passed on.

*India did not fail to invent. India invented the technologies that cannot be seen, cannot be owned, and cannot be destroyed. And gave them, freely, to all of humanity.*

## What Cannot Be Built Outside Must Be Built Within

Look at what the three technologies of this chapter have in common.

Mantra operates through vibration — but the receiver of the vibration is the human nervous system. Without a sufficiently quiet nervous system, the Mantra is just sound. Yantra operates through visual focus — but the focus is the human mind. Without a sufficiently still mind, the Yantra is just a drawing. Tantra operates through the field of consciousness — but the field is entered through the human being. Without a sufficiently open human being, the network is closed.

**Every one of India's invisible technologies depends on the same instrument. The human being. The body. The mind. The breath. The receiver.**

And this raises the question that the next chapter takes up directly. If the receiver is the technology — if all of India's other inventions are extensions of a single fundamental instrument that every human being carries — then what, exactly, is the receiver? How does it work? And what does it mean to live in a universe where the human being is not the passive observer of an external world but an active participant in a field that is, in some sense, recording everything?

The next chapter goes to the foundation. To the verse that Vagbhata placed at the opening of the Ashtanga Hridayam (Heart of the Eight Branches of Medicine), to a Mahavakya (great saying) from the Upanishads (philosophical texts) that contains the most radical claim any civilization has ever made about the nature of the self, and to a piece of science the modern world is only now discovering — that the universe is, in the most literal sense, a recording device. And the receiver of that recording is the one technology India spent five thousand years refining: you.

CHAPTER EIGHT

# The Receiver

*What Is Already Within*

अहम् ब्रह्म अस्मि।

*I am Brahman. I am the universal consciousness — the same field from which Brahma himself remembered.*

*— Brihadaranyaka Upanishad 1.4.10 (Yajur Veda) — one of the four Mahavakyas (great sayings) of the Upanishads*

## The Question Inside Brahma Smritva

There is a verse at the very beginning of the Ashtanga Hridayam (Heart of the Eight Branches of Medicine) — Vagbhata's foundational text of Ayurveda (ancient Indian health science) — that contains a single word strange enough to make you stop reading.

The verse describes how the knowledge of Ayurveda came into the world. It tells us that Brahma — the creator of the universe — transmitted Ayurveda to Daksha Prajapati, who transmitted it to the Ashwini Kumars (divine physicians), who transmitted it to Indra (king of the gods), who transmitted it to Atreya and the great sages, who transmitted it to Agnivesha and other disciples, who composed their own separate treatises on the subject.

This much is ordinary lineage. Knowledge is passed down from teacher to student, as knowledge always is.

But the verse does not say Brahma invented Ayurveda. It does not say Brahma discovered it. It does not say Brahma composed it.

**It uses one specific Sanskrit (ancient Indian language) word to describe what Brahma did: Smritva. Brahma remembered.**

Read that again. The creator of the universe — the source of all manifestation — did not invent the knowledge of life. He remembered it.

If even Brahma remembered, then the knowledge existed before Brahma. Somewhere. In some form. Waiting to be remembered.

Where?

## Where Knowledge Actually Lives

The modern world has a theory about where knowledge lives. It is so deeply embedded in education systems, in research institutions, in the structure of universities and laboratories, that almost nobody examines it. The theory is this: knowledge does not exist until a human mind creates it. The universe is, in its natural state, unknown. Knowledge is built — by observing, hypothesizing, experimenting, and accumulating evidence — piece by piece, generation by generation.

This is the operating assumption of modernity. And it has produced extraordinary external results — the inventions and machines and systems we have already discussed in this book.

But there is a second theory. India's theory. And the second theory says something that, on first hearing, sounds impossible:

*Knowledge already exists. It has always existed. It exists in its complete form in the fabric of consciousness itself — in what India called Brahman, the universal field that underlies all being. The task of the human mind is not to create knowledge. The task is to become quiet enough to receive what is already there.*

If this second theory is correct, then the entire architecture of modern education — built around the idea that human beings begin empty and must be filled — is solving the wrong problem. The human being is not empty. The human being is already a receiver, tuned to a field that contains everything, but unable to hear the signal because of the noise of Raag (desire/attachment) — the noise this book has been examining from its first page.

## Aham Brahmasmi — The Most Radical Statement in Human History

There are four great sayings in the Upanishads (ancient philosophical texts) — the Mahavakyas. Each one, in just a few syllables, contains a complete philosophy.

One of these four sayings is Aham Brahmasmi (I am Brahman). It appears in the Brihadaranyaka Upanishad — one of the oldest surviving Indian philosophical texts. And what it says is this:

**The 'I' that is reading these words is not separate from the universe. The 'I' is the universe, momentarily looking through one set of eyes.**

Now this can sound, to the modern ear, like inflated mysticism. It is anything but. The 'I' in Aham Brahmasmi does not refer to the ordinary ego — the small self that has a name and a job and a list of problems. The 'I' refers to the witnessing awareness that exists before any of those things. The consciousness that observes the thoughts but is not itself a thought. The presence that sees through the eyes but is not the eyes.

That witnessing consciousness, says the Upanishad, is identical with the universal consciousness — Brahman. Not similar to it. Identical.

If this is true, then something extraordinary follows. The same field of consciousness from which Brahma remembered Ayurveda is the field of consciousness that you, right now, are reading these words from. You have access to the same source. You always have. The barrier between you and the universal knowledge is not distance. It is not technology. It is not vocabulary. The barrier is noise.

*Brahma remembered. So can you. The only difference is the noise.*

## The Universe Is Recording

Before we examine what it means that consciousness is a receiver, we have to look at something modern science has discovered that no Indian sage of two thousand years ago could have known in scientific terms — and that nevertheless confirms, with mathematical precision, what those sages were saying.

**The universe is, factually and demonstrably, a recording device.**

This is not a poetic claim. It is a literal statement supported by multiple independent scientific findings. Let us look at them.

## The First Recording — The Cosmic Microwave Background

In 1964, two scientists working on a radio antenna detected a faint hum coming from every direction in the sky. They could not eliminate it. At first they thought it was equipment noise. It was not. The hum was the radiation left over from the early moments of the universe itself — the cooled remnant of the first light that could ever travel freely through space, emitted approximately 380,000 years after the universe began. It is now called the Cosmic Microwave Background, and it fills the entire universe in every direction.

Think about what this means. The universe contains, right now, an actual physical recording of itself from nearly 14 billion years ago. That recording is still being received. Every moment. By any instrument sensitive enough to detect it. The information has not gone away. It cannot go away. It is the radiation of the universe's own birth, still traveling, still arriving.

## The Second Recording — Every Human Voice Ever Broadcast

Since the first radio transmission in 1895, every signal sent through a radio tower, every television broadcast, every wireless transmission has been propagating outward through space at the speed of light. They become weaker with distance — the signal strength falls as the square of the distance from the source — but they do not stop existing. They cannot. The laws of physics do not permit electromagnetic radiation to simply vanish.

This means that, in a very precise sense, the universe is currently carrying a sphere of human voices and human music and human speech approximately 250 light-years across, and growing every second. A sufficiently sensitive receiver — at the right distance, in the right direction — could, in principle, hear a radio broadcast made on earth a hundred years ago. The broadcast is still there. It has just moved further away.

## The Third Recording — The Earth Writes Its Own Diary

Long before any human invented radio or built a satellite, the earth was already keeping its own continuous record. Tree rings record every year of weather — wet years, dry years, fire years, volcanic years — for the entire life of the tree. Ice cores drilled from glaciers in Antarctica and Greenland hold preserved samples of the earth's atmosphere going back 800,000 years. Deep ocean sediment cores carry records going back hundreds of millions of years. The earth has been writing its diary in trees, ice, mud, and stone since long before any human being arrived to read it.

Modern science can now read these natural recordings with extraordinary precision. We know what the climate was like during the time of the dinosaurs. We know what the atmosphere contained during the last ice age. We know which volcanic eruptions disrupted which civilizations. The earth recorded all of it. It simply waited for someone to learn how to read what it had written.

## The Fourth Recording — Your Own DNA

Every cell in your body contains a molecule — DNA (the biological blueprint molecule) — that records the entire evolutionary history of life on earth. Three billion years of trial and error, of adaptation and survival, of every ancestor who ever lived and managed to reproduce. The information is in you right now. Encoded. Compressed. Folded into a structure so dense that the DNA from a single one of your cells, if fully uncoiled, would stretch to approximately two meters in length.

And the field of Epigenetics (the study of how environmental factors and experiences affect gene expression) has revealed that the way our genes are expressed can be shaped by the experiences of previous generations — that the body remembers more than just biological code. The chemical markings on DNA appear to carry traces of what one's ancestors lived through. The body is a recording device. It has been recording for as long as there have been bodies.

## The Fifth Recording — The Holographic Principle

This last finding is, even by the standards of modern physics, strange. Some of the most respected theoretical physicists working today — including Nobel laureate Gerard 't Hooft and Leonard Susskind — have proposed what is now called the Holographic Principle. It suggests that the universe, or at least major regions of it, can be described as a hologram — that all the information encoded within a three-dimensional region of space is actually stored on a lower-dimensional boundary of that region.

The mathematics is more rigorous than the metaphor sounds. Studies have suggested that the maximum amount of information any region of space can contain is limited by the surface area of that region, not its volume. The implication, if it holds, is profound: information may be the fundamental fabric of reality. The universe may be, at its deepest level, a structured field of information — exactly the kind of substrate from which knowledge could, in principle, be received.

**Cosmic Microwave Background. Radio waves still traveling. Tree rings. Ice cores. DNA. The Holographic Principle. By five completely independent lines of evidence, modern science is converging on a single conclusion — the universe records. It does not forget.**

### If the Universe Records — What Is the Receiver?

Here is where the modern scientific finding and the ancient Indian understanding meet.

If the universe is, in fact, a recording — if information is the fundamental fabric of reality — then a question arises that modern science has not yet known how to answer. Every recording requires a receiver. The Cosmic Microwave Background was discovered when a radio antenna detected it. Tree rings tell their story only when a researcher cores the tree and counts the rings. DNA's information becomes accessible only when sequencing instruments decode it. The universe records, but the recording becomes knowledge only when something receives it.

Modern science has built increasingly sensitive instruments to receive these recordings. Telescopes that detect radiation from the edges of space. Particle accelerators that probe the fabric of matter. Genetic sequencers that read the code in cells.

India asked a different question. Not — what instruments can we build to receive the recording? But — what instrument are we already?

**The human being, says the Indian understanding, is the most sensitive receiver in the universe. The brain is the most sophisticated signal-processing system known to exist. Consciousness is the medium through which information becomes knowledge.**

If you were a sufficiently advanced civilization, you would not build instruments external to yourself to receive the universe's information. You would refine the instrument you already are. You would learn to quiet the noise. You would develop the technology of stillness.

This is exactly what India did. For five thousand years.

## The Lemon — A Small Demonstration

Before we go further into the universal claim, let us look at something small and immediate. Something you can verify in your own body within the next thirty seconds.

Stop reading for a moment. Close your eyes. Imagine that you are holding a fresh lemon. Imagine the weight of it in your hand. Imagine cutting it open — the bright yellow flesh, the juice running over your fingers, the sharp citrus smell. Now imagine biting into it. Imagine the sour juice flooding your tongue.

Now notice your mouth.

**It is watering. There is no lemon. There has never been a lemon. The thought of a lemon has caused real, physical change in your body.**

Saliva is being produced. Glands are activating. Your body is responding to a fruit that does not exist, has never existed in this room, and never will. The boundary between what is only thought and what is real — the boundary the modern world treats as solid — turns out to be one of the thinnest and most permeable boundaries in nature.

This single, simple, repeatable observation contains a profound implication. The mind is not separate from the body. Thought is not separate from physical effect. What happens in the imagination has measurable consequences in the flesh. And the human being — the body and mind together — is not the passive observer of an external world. It is a participant. A receiver. A device that responds to what is imagined as if the imagined thing were real.

## Thought Becomes Body — The Wider Picture

The lemon experiment is one small demonstration of a much wider phenomenon. The body responds to thought constantly. Not metaphorically. Physiologically.

Stress response — imagine a difficult conversation with someone you fear. Notice what happens. Your heart rate rises. Cortisol — the body's stress hormone — increases in your blood. Your muscles tighten. The conversation has not happened. The fear is imagined. The body's response is real.

Grief — the memory of a loved one who has died can produce tears years after their death. No external event has occurred. A neural pattern has activated. Tears arrive in the present from a thought about the past.

Placebo and nocebo effects — a patient given a sugar pill they believe is medicine often heals — measurably, in controlled studies. A patient given a sugar pill they believe will harm them often becomes sicker. The pill is chemically inert in both cases. Belief alone produces physical change.

Phantom limb — a person whose limb has been amputated continues to feel pain in the limb that no longer exists. The pain is real. The limb is not. The brain is generating the sensation in the absence of the body part.

Ayurveda's Swapna-dosha — the traditional Ayurvedic (ancient Indian health science) tradition has long described a particular phenomenon in young men, usually between the ages of fourteen and sixteen — though with significant individual variation — in which physical sexual response occurs during sleep, in response to mental imagery alone. The tradition records this without surprise and without judgment, because the underlying principle was already understood: the body does not distinguish, in any clean way, between what is happening in the outside world and what is happening in the inner world of thought and image.

Modern neuroscience confirms what all of these examples demonstrate. The same brain regions activate whether you actually perform an action or vividly imagine performing it. The body responds to thought as if thought were event. This is why mental rehearsal — practiced by athletes, musicians, surgeons — produces measurable improvements in physical performance. Imagined practice activates the same neural pathways as physical practice. The brain does not, at the deepest level, distinguish.

**If thought can produce physical change in the body that thinks the thought — what is thought, really? And if the boundary between imagined and real is so permeable inside one body, what does that say about the boundary between one mind and the universe in which it lives?**

## When a Friend Appears

You have experienced this. Almost everyone has.

You are sitting with a friend. You begin talking about another person — someone who has no particular reason to appear at this moment in your life. They are not expected. They have not called. They live in another city, perhaps another country. And within minutes — sometimes within seconds — that person calls, or messages, or walks through the door.

In India, the traditional response to such a coincidence is to say to the person who has appeared — you will live for a hundred years. The cultural meaning is affectionate, but the underlying observation is precise:

**The person did not appear randomly. They appeared because they were already in motion. Their consciousness was already moving toward you. Their thoughts had already turned in your direction. And that movement of attention — that broadcast through the field that India calls Tantra (the universal network) — reached you and produced, in your mind, the impulse to speak of them.**

The signal arrived before the physical arrival. Their footsteps were the confirmation, not the cause.

If the universe is a recording — if information moves through the field of consciousness the way radio waves move through space — then this small, everyday observation begins to make a different kind of sense. The Karma (action and intention) of the friend who was thinking about you was already a broadcast. Your stillness — the quiet moment of conversation in which the thought of them surfaced — was the reception.

This is not magic. This is, in the Indian understanding, the natural operation of consciousness in a field that contains everything. The receiver is always on. The only question is whether the noise is quiet enough for the signal to be heard.

## Ek Patta Bhi Nahi Hilta — Not a Single Leaf Moves Without Reason

There is a phrase in Hindi (the language of much of north India) that captures something the Indian mind has carried for thousands of years.

*Ek patta bhi nahi hilta. Not a single leaf moves without reason.*

The phrase is usually spoken with quiet certainty, sometimes with resignation, sometimes with peace. It is not fatalism in the gloomy sense. It is the recognition that the universe is, in some way that the modern mind has not learned how to think about, already complete. Already coherent. Already moving according to a pattern that human beings can sometimes glimpse but not author.

In the language of this chapter — if the universe records everything, if consciousness is the receiver, if information is the fabric of reality — then it follows that every event is part of a larger pattern. The leaf that moves on a tree does so because of wind. The wind blows because of pressure differences. The pressure differences exist because of solar heating. The solar heating reaches that exact tree at that exact moment because of an unbroken chain of causes going back to the first moments of the universe.

This is not Indian philosophy alone. This is causality, as modern physics describes it. Every effect has a cause. Every cause has its own cause. Trace any single event back far enough, and you arrive at the beginning of the universe itself.

**Not a single leaf moves without reason. India said this. Physics confirms it. The question is only — who can read the script?**

This is where the Indian and modern understandings diverge. Modern physics says the script exists but is too complex to read in any meaningful way. The Indian understanding says the script exists, the script is being broadcast continuously through the field of consciousness, and certain instruments — certain human beings, when sufficiently quiet — can read pieces of it.

## Saraswati on Manthara's Tongue

There is a story in the Ramayana (ancient epic of Rama) — particularly in the Ramcharitmanas (the Hindi retelling of the Ramayana by the poet Tulsidas) — that, when read carefully, contains the entire argument of this chapter compressed into a single dramatic moment.

The setting: Ayodhya. Lord Rama is about to be crowned king. The kingdom is celebrating. Everything is in place. The most beloved prince in the history of the dynasty is one day away from his rightful throne.

And then, in a single night, everything is undone.

Manthara — a hunchbacked maidservant in the palace, of no particular consequence in the political hierarchy — speaks to Queen Kaikeyi, Rama's stepmother. Manthara fills Kaikeyi's heart with suspicion. She poisons Kaikeyi against Rama. By the end of the conversation, Kaikeyi has demanded that Rama be exiled to the forest for fourteen years and that her own son, Bharat, be crowned king instead.

On the surface, this looks like the simplest kind of palace drama. A jealous queen, a manipulative servant, an unjust outcome.

But Tulsidas, writing the Ramcharitmanas, includes a detail that transforms the entire story.

**Tulsidas writes that Saraswati — the goddess of speech and wisdom — sat herself upon Manthara's tongue and put the poisonous words into her mouth. Not metaphorically. Deliberately. As a cosmic act.**

The poison that exiled Rama did not come from Manthara's small mind. It came from the goddess of wisdom herself, acting in service of a script far larger than any human being in that palace could see.

Why? Because Rama's exile was the script. Without the exile, Rama does not encounter the forest. Without the forest, he does not meet the demons that test him. Without the demons, Sita is not abducted. Without Sita's abduction, Ravana is not defeated. Without Ravana's defeat, the demonstration of Dharma (right action) that the whole Avatar (divine descent) was meant to make does not occur.

Saraswati — the goddess of wisdom — knew the script. And so she sat on the tongue of an apparently malicious maidservant and ensured that the next scene began on time.

*Even the cruelty that exiled Rama was, at the level of the cosmic script, an act of divine wisdom. Even Manthara was, for that one night, an instrument of Saraswati.*

This is what the Indian tradition means when it says ek patta bhi nahi hilta. Not that human beings have no choices. Not that suffering does not matter. But that there is a larger pattern moving through every apparent accident, every apparent cruelty, every apparent setback. And the human being who is quiet enough to listen begins, sometimes, to hear the pattern.

## Ram and Krishna — Two Sides of the Same Scale

If Ram was the rule-follower in his Avatar, Krishna was the rule-breaker in his. Both serve the same script. They simply play opposite roles, in opposite ages, by opposite methods.

Rama is described in the Ramayana as Maryada Purushottam — the perfect upholder of moral and social order. He follows his father's word even when his father's word is unjust. He accepts exile even when the exile is the result of a manipulation. He observes every rule of his role as son, as husband, as king. His Dharma is expressed through obedience to the established order.

Krishna is the opposite. Krishna lies. Krishna deceives. Krishna engineers situations in which the rules of warfare are quietly broken so that the war can be won by the side of Dharma. In the Mahabharata (ancient epic), Krishna persuades Arjuna to fire on Karna while Karna is unarmed and trying to free his chariot wheel. Krishna engineers the death of Drona by telling a half-truth that Yudhisthira repeats. Krishna is honest about his methods — and his methods are, by any conventional measure, deceitful.

**Two Avatars. Two opposite methods. And both are the same God.**

This is one of the deepest insights in Indian philosophy. The universe is not a moral system in which one rule applies in all situations. The universe is a Taraaju — a balancing scale. And what restores balance in one age is exactly the opposite of what restores balance in another.

The Treta Yuga — Rama's age — was a time when Dharma still mostly held. The rules still worked. What was needed was a perfect demonstration of how to live within the rules. So the Avatar of that age was Rama — the one who followed every rule without exception.

The Dwapara Yuga — Krishna's age — was a time when Dharma had been broken so thoroughly by the powerful that following the rules would have meant letting the rule-breakers win. What was needed was someone willing to break the rules in service of a deeper Dharma. So the Avatar of that age was Krishna — the one who broke the rules so that the principle the rules existed to protect could survive.

*Krishna means black. Krishna's color is black because black contains all colors. Krishna's role is balance because balance — true balance — requires holding opposites in one being.*

There is a phrase in Indian thought that captures this — that a small amount of strategic untruth, used carefully and in service of Dharma, is like salt in dough. It does not dominate the dough. It is invisible in the final bread. But without it, the bread has no taste. Without it, the larger structure does not work.

**Rama is honesty. Krishna is balance. Both serve the same cosmic script. The script needs both. And the human being who is quiet enough to read the script begins to understand when to be Rama, and when to be Krishna.**

## The Vedas Were Heard, Not Composed

There is one more piece of evidence in the Indian tradition for everything this chapter has argued. And it has been hiding in a single word for thousands of years.

The Vedas (the oldest Indian sacred texts) — the foundational texts from which all subsequent Indian philosophy descends — are not called writings. They are not called compositions. They are not called inventions of the sages who recorded them.

**They are called Shruti. Sanskrit for that which is heard.**

The Rishis (sages/seers) who gave us the Vedas did not, according to the tradition, sit down and compose them. They sat in deep silence — for years, sometimes for lifetimes — and they heard them. The knowledge arrived. It did not have to be invented because it was already there, broadcasting through the field of consciousness, waiting for an instrument quiet enough to receive it.

Read with the framework of this chapter, this becomes a precise and astonishing claim. The Rishis were claiming, in advance of any modern understanding of electromagnetic fields or information theory or the recording function of the universe, that knowledge is a signal that can be received. That the human being is the receiver. That the technology required to operate the receiver is not external — it is internal. It is stillness. It is the absence of Raag. It is the willingness to be empty enough that the signal can be heard.

This is not a religious claim that requires faith. This is a technological claim that requires practice. India developed the practice. The practice is what we have been calling, throughout this book, Yoga (mind-body union), Pranayama (breath regulation), Dhyana (meditation), Mantra (sacred sound). All of these are, in the language of this chapter, instruments for tuning the receiver.

## Why Stillness Was India's Technology

Now everything in this book reorganizes itself.

The white clothing of Chapter Two was not simple cloth. It was a daily reduction of Raag — a tuning of the receiver to a lower noise floor.

The Mahalakshmi worship of Chapter Three was not abstract piety. It was the placing of Buddhi (intelligence) before Dravya (material wealth) so that the receiver would not be drowned by the noise of accumulation.

The cow of Chapter Five was not livestock. It was a living model of a complete system, contemplated daily by a civilization training itself to perceive completeness.

The Pranayama and Dhyana of Chapter Six and Chapter Seven were not exercises. They were the operating instructions for the internal instrument.

All of it. The entire civilizational architecture. Was a sustained, multi-millennial effort to do one thing — keep the receiver clean enough that the signal could be heard.

**The modern world is building larger and larger external instruments to receive the universe's signals. India built the most refined internal instrument and gave the operating manual to anyone willing to practice.**

And here is the quietest insight of this chapter, the one that will sit beneath everything that follows:

*The receiver was never broken. The receiver was never absent. The receiver is what you are. The signal has never stopped broadcasting. The only thing that has ever needed adjustment is the noise.*

Reduce the noise — reduce the Raag (desire) — and what is already within rises into awareness. Not as an achievement. Not as an invention. As a remembering.

## What This Chapter Has Set Up

Before we move to the next chapter, let us state plainly what this chapter has established.

First — the universe is a recording. Modern science confirms this through at least five independent lines of evidence.

Second — the human being is a receiver. The body responds to thought as physical event. The mind perceives information beyond its sensory inputs. Consciousness is the medium through which the universe's recording becomes knowledge.

Third — Brahma did not invent Ayurveda. He remembered it. The lineage from Brahma down to the human sages was not a chain of inventions but a chain of receptions, each one quieter than the last so that the original signal could be transmitted without distortion.

Fourth — Aham Brahmasmi is not poetic exaggeration. It is the precise description of what the witnessing consciousness inside every human being actually is — a localized expression of the universal field from which Brahma himself remembered.

Fifth — Ek patta bhi nahi hilta. Not a single leaf moves without reason. The cosmic script is being written continuously, broadcast continuously, and can be read — partially, imperfectly, but really — by any consciousness still enough to listen.

Sixth — Saraswati on Manthara's tongue, Ram and Krishna as opposite Avatars of the same Dharma, the Vedas as Shruti rather than composition: the Indian tradition has been telling us, in many forms, for many thousands of years, that the script is real and the receiver is us.

**And if all of this is so — if every human being is, by simply being human, a receiver tuned to the universal field — then a question arises that the next chapter will answer.**

*If knowledge is already within, what else is already within? What about the body's wisdom about how to heal itself? What about the mind's understanding of who it really is? What about the heart's natural orientation toward what is true?*

The chapter that follows takes this question to its conclusion. It argues that every human being is born not only with the receiver, but with the complete signal. That the newborn child is, in the most literal sense, the universe momentarily condensed into a small body. That the project of a human life is not addition but subtraction — not gaining what is missing but releasing the noise that obscures what is already there.

*In the next chapter, we will see this most radical of all Indian ideas demonstrated in three lives: a newborn child in any mother's arms anywhere on earth, a tribal student named Eklavya practicing archery before a clay statue in a forest, and a great warrior named Arjuna who, on the battlefield of Kurukshetra, was reminded by Krishna of what he had always already known.*

CHAPTER NINE

# Born Complete

*The Living Proof of the Receiver*

## The Living Proof

In the previous chapter, we examined a claim that the Indian tradition has been making for thousands of years — that every human being is a receiver, tuned to a universal field of consciousness, capable of accessing what was always already within. The argument was philosophical, supported by science, framed by Brahma Smritva (Brahma's remembering) and Aham Brahmasmi (I am the universal).

All of this could be dismissed as theory. The mind can easily reject what it cannot see.

So the next question is the right one to ask. If every human being is born as a complete receiver of the universal field — if Aham Brahmasmi (I am that universal consciousness) is true at the deepest level — then where is the proof? Where is the evidence that any of this actually shows up in real human life?

**The proof is lying in the arms of every new mother right now, in every corner of the world.**

The newborn child. Not as a symbol. As the most ordinary, common, repeatable miracle on earth.

A baby in the first weeks of life survives entirely on mother's milk. Not on solid food. Not on water — many traditional Indian practices, in alignment with current World Health Organization recommendations for exclusive breastfeeding in the early months, hold that the newborn should not be given water before solid food is introduced, because mother's milk contains all the hydration the infant needs. The infant does not need anything beyond what it receives from the mother.

**It is, in every biological sense, complete.**

And the consciousness of that newborn — the awareness that inhabits the tiny body in those first weeks — is the purest form of Purnam (completeness) that a human being will ever be in this lifetime. There is no Raag (desire/attachment) yet. There is no Kama (desire for pleasure), no Krodha (anger), no Lobha (greed), no Matsarya (jealousy). There is only pure being. Pure presence. Pure awareness without content.

This is what Aham Brahmasmi looks like in a small human body. It is what the receiver of the previous chapter looks like before noise has entered the signal. And it is something every reader of this book has been. Every one of us, without exception, began as this living, breathing, breathing proof of the central argument.

## What Purnam (Wholeness) Really Means

At the very beginning of this book — before the Preface, before Chapter One, on a page that stands alone in silence — there is a Shloka (verse) from the Isha Upanishad (ancient Indian philosophical text):

ओं पूर्णमदः पूर्णमिदं पूर्णात् पूर्णमुदच्यते । पूर्णस्य पूर्णमादाय पूर्णमेवावशिष्यते ॥

Purnam (complete/whole) does not mean perfect in the way the word is commonly used. Purnam means whole. Complete. Not lacking anything. The invisible universe is Purnam (complete). The visible world is Purnam. And what emerges from completeness is always complete. Even when you take completeness away from completeness — what remains is still complete.

**And you — a human being, born into this world — are also Purnam. Not lacking anything essential. Not missing any knowledge that you fundamentally need. Born with the full inheritance of the universe within you.**

The problem is not that you lack knowledge. The problem is that you are unconscious of what you already have. As the previous chapter argued — the receiver is in place. The signal is being broadcast. The interference of Raag (desire) is what stands between you and your own completeness.

## Where Raag Enters the Newborn

Watch what happens to the complete newborn. The child begins to eat. Food enters the equation. And with food — Raag begins.

The child wants milk and cries when it is not there. This is the first desire. The first attachment to an external object. The first moment when the consciousness that was Purnam begins to experience itself as needing something outside itself. The first moment of Raag.

This is the deepest connection between Chapter One and this chapter. Raag — the root disease identified in the first verse of the Ashtanga Hridayam (Heart of the Eight Branches) — begins exactly where Born Complete ends. The newborn is Purnam. Food begins. Raag begins. And from that moment, the journey of this entire book begins — the long, beautiful, difficult journey back to what we were before food arrived.

**The receiver was clear. Then desire entered. The receiver did not break. It only became harder to hear.**

## The Universe Imprinted at Birth

Indian Jyotisha (astrology — the science of light and planetary positions) holds something extraordinary about the moment of birth.

At the exact moment a child is born — at that precise instant when it takes its first breath and enters the electromagnetic field (energy field) of the earth independently — the position of every planet in the solar system is, according to this tradition, imprinted on the child's consciousness. Not just the planets. The Rishis (sages) understood this imprinting to extend to the position of every significant mass of matter in the cosmos, to the energetic state of every living and non-living thing in the universe at that exact moment.

This is the Janma Kundali (birth chart) — the map of the universe at the moment of one's birth. Every Indian child has one. Every significant life decision — marriage, career, home, travel — has traditionally been considered in relationship to this map.

But look at what this system is actually claiming, through the lens of this chapter and the previous one:

*Every human being is born with the complete picture of the universe at the moment of their arrival encoded in their consciousness. They carry a map of where every planet, every significant energy, every cosmic force was positioned at the instant of their birth. They are, from the first breath, a hologram of the universe at that moment.*

**This is the most literal expression of Born Complete that Indian civilization ever produced. You did not come into the world empty**

**and waiting to be filled with knowledge. You came in carrying a complete map of the universe at the moment of your arrival. The Jyotisha (astrology) tradition exists to help you read that map — not to impose a destiny on you, but to help you understand the particular configuration of consciousness you arrived with.**

## The Sutika Period — India's Original NICU

Modern medicine has a NICU — the Neonatal Intensive Care Unit (specialized care ward for newborns). It is a sealed environment. Controlled temperature. Controlled light. Controlled sound. Minimal visitors. Maximum rest for the newborn, who has just completed the most demanding physical journey of its existence — the journey from womb to world.

India had this. For thousands of years. For every birth. Not just for premature or vulnerable babies — for every child, and equally for every mother.

It was called the Sutika period (postpartum care period) — sometimes called Jaappa in different regions of India. The traditional practice prescribes approximately forty days of complete rest, isolation, and intensive care for the mother and newborn after birth. No visitors — or strictly limited, close family only. No exposure to the general environment. Specific foods. Specific practices. Complete protection of the mother's recovering body and the newborn's emerging consciousness.

The Ayurvedic (ancient Indian health science) prescriptions for this period are extraordinarily detailed. Ajwain (carom seeds — a warming spice with digestive and anti-inflammatory properties) preparations are given to the mother — in various forms depending on the region: as a drink, as a poultice, as a component of the first solid foods after birth. The warming properties of Ajwain help restore the Agni (digestive fire) that is depleted during childbirth. The anti-inflammatory properties support the healing of the birth canal. The carminative (gas-reducing) properties help manage the digestive disruption of the postpartum period.

Warm Ghee (clarified butter). Specific Dals (lentil preparations) chosen for their ease of digestion and nutritional density. Methi (fenugreek) preparations to support milk production. Sesame-based foods for warmth and mineral replenishment. Each ingredient chosen with specific intention for the specific physiological state of a woman who has just given birth.

These practices vary across India's different regions and communities — each variation reflecting the specific knowledge of the local ecology, the specific foods available, the specific climate to be managed. But the underlying principle is consistent everywhere: the first forty days after birth are sacred. The mother and child are in a state of extraordinary vulnerability and extraordinary potential simultaneously. They must be protected, nourished, and allowed to establish their connection in peace.

A growing body of postpartum research suggests what India long understood: the weeks after birth are a period of profound neurological, hormonal, and psychological reorganization. The quality of support a mother receives in this period appears to shape her long-term wellbeing and the quality of the attachment bond with her child.

**The Sutika period was India's answer — developed across thousands of years of observation, encoded in community practice, passed from mother to daughter across generations without any institution, without any credential, without any cost.**

## Three Days of Rest

The Sutika period has a smaller, monthly sister — a practice still alive in households across India, including the one I grew up in.

For three days of a woman's monthly cycle, in many Indian communities, she does not cook. She does not enter the kitchen. She cannot be asked to fetch, to serve, to carry. The household adjusts around her — because the rule is not a preference. It is a rule.

The modern world has often looked at this practice and seen only exclusion. But consider what the rule actually accomplishes, in a world where a woman's domestic labor otherwise never stops. Ayurveda describes the days of menstruation as days when the body is engaged in demanding internal work — blood is being lost, the downward-moving vital energy is fully occupied, and the classical texts prescribe exactly what the household rule enforces: rest, light food, freedom from physical strain. A recommendation of care can be skipped, especially by the woman herself, who will always find one more task that needs doing. A ritual rule cannot be skipped. India made the rest non-negotiable by encoding it as ritual.

The modern world arrived at menstrual leave as a policy debate in the twenty-first century. India delivered it as a household rule thousands of years earlier — transmitted so effectively that a woman who never read a single text enforces it with precision, because her mother did, and her mother's mother before that.

The same protective principle. Two scales. Forty days after a birth. Three days every month. Both encoded, both delivered, both carried by the same unbroken chain of women.

## Brahma, Vishnu, Mahesh — Within Every Human

In Chapter Three, we introduced the Trimurti (holy trinity) — Brahma (creator), Vishnu (preserver), and Mahesh/Shiva (transformer/dissolver) — as the three fundamental forces that govern all of existence.

Now we can understand their significance in the context of Born Complete.

**Every human being contains all three. Not as a metaphor. As a functional reality.**

The Brahma within — the creative impulse. The part of you that generates new ideas, new solutions, new possibilities. The impulse to begin. When you are free from the Raag of performance anxiety, this creative force flows naturally and endlessly.

The Vishnu within — the sustaining principle. The part of you that maintains what is valuable, that shows up consistently, that builds the trust and reliability on which all genuine relationships and genuine achievements depend.

The Mahesh/Shiva (transformer) within — the dissolving and transforming force. The part of you that can release what no longer serves, that can let go of the comfortable but diminishing, that can give the thumb when the examination asks for it.

You were born with all three. The entire creative, preserving, and transforming capacity of the universe is within you. Raag distorts all three. It distorts creation into ego-driven accumulation. It distorts preservation into fearful hoarding. It distorts transformation into destruction rather than release.

**When Raag is reduced — when the interference is cleared — all three operate as they were designed to. Creation flows. Preservation**

**builds. Transformation releases. And the human being who contains all three in their natural state is a complete civilizational force in a single body.**

## Yoga — The Path Back to Completeness

The word Yoga (union) comes from the Sanskrit (ancient Indian language) root Yuj — meaning to join, to unite, to yoke. Yoga does not mean flexibility. It does not mean exercise. It means union — the reuniting of the individual consciousness with the universal consciousness it was never truly separated from.

**Yoga is not the achievement of something new. It is the return to what was always already complete.**

And Yoga is not one path. The ancient Indian understanding recognizes four primary paths — each one suited to a different dimension of the human being, each one a different route to the same destination:

Karma Yoga (yoga of selfless action) — the path of the body, of service, of action without ego and without attachment to outcome. The Karma Yogi (practitioner of selfless action) dissolves Raag through the act of complete service — giving everything, holding nothing back, asking for nothing in return. This is Eklavya in the forest. This is the woman cleaning her child's waste without any thought of recognition.

Bhakti Yoga (yoga of devotion) — the path of the emotions, of pure love, of devotion that has been completely freed from desire. The Bhakta (devotee) uses the same emotional energy that Raag corrupts into Moha (attachment) and transforms it into unconditional love. Eklavya's devotion to the clay statue of Dronacharya was Bhakti Yoga (devotional path) in its purest form.

Gyana Yoga (yoga of knowledge) — the path of the mind and intellect, of deep inquiry, of the investigation of the nature of reality until the one who is inquiring and the one being inquired about are seen to be the same. The Gyana Yogi (practitioner of knowledge path) does not accumulate knowledge — they use inquiry to dissolve the illusion of separation.

Kriya Yoga (yoga of energy) — the path of working directly with Prana (life force), the internal electrical system we explored in Chapter Six. The Kriya Yogi (energy practitioner) uses specific practices — breath, sound, movement, specific techniques for directing Prana — to clear the Nadi (energy channel) network and allow the natural completeness of consciousness to emerge.

**Four paths. One destination. Not achievement. Not arrival at something new. Reunion. With what was always complete.**

## What Modern Science Is Finding

The claim that every human being is born with complete knowledge sounds, to the modern ear, like mysticism. So let us look at what modern science has been discovering about what is encoded within each human being at birth.

A human being begins as a single cell — a fertilized egg approximately 0.1 millimeters in diameter. Within that single cell is DNA (the biological blueprint molecule) that, if fully uncoiled, would stretch to approximately two meters in length. This molecule contains the complete instructions for building an entire human body — encoded in a chemical language of four letters arranged in sequences of approximately three billion base pairs.

The field of Epigenetics (the study of how environmental factors and experiences affect gene expression) has revealed that the way our genes are expressed can be shaped by the experiences of previous generations. Some research has explored how stress responses and certain adaptive behaviors of parents and grandparents may be reflected in the chemical markings on their descendants' DNA. The full extent and mechanisms of this transgenerational inheritance in humans remain an active area of investigation, but the basic finding — that we inherit more than just our genetic code — is increasingly accepted.

And then there are the documented cases of savants — people who, following brain injury or in certain developmental conditions, suddenly access abilities that are very difficult to explain by ordinary learning or practice. There are well-documented cases of individuals who, after head injuries, have woken up able to play complex classical piano pieces they had never trained for, or to draw architecturally precise buildings from a single glance. These cases are rare and the mechanisms are debated — but they exist, in the medical literature, and they raise a question that has not been fully answered:

Where does this knowledge come from? It does not appear to have been learned in the conventional sense. It emerges — as if it was already there, waiting for the right conditions to become accessible.

**Modern psychology has long described a shared layer of human experience that contains universal symbols, universal myths, and universal patterns appearing independently across every culture. India called this layer Brahman (universal consciousness). Different vocabularies, pointing toward related territory.**

## The Obstacle Is Raag

If every human being is born with the complete knowledge of the universe — why does almost nobody experience this completeness? Why do we spend our entire lives searching for knowledge, accumulating information, building understanding piece by piece from the outside in?

**The answer, India says, is Raag.**

Raag — desire, attachment, ego, fear, comparison — creates a constant interference pattern that obscures the signal of the deeper knowledge within. Think of it the way the previous chapter described it. The receiver is in place. The signal is being broadcast. But if the receiver is full of static — if the mind is full of desire and fear and comparison and ego — you cannot hear the signal of your own completeness. The station you are trying to reach is still broadcasting. You cannot hear it through the interference.

Raag is the interference. And in the age of artificial intelligence — where every human being now carries access to more information than any previous civilization possessed — the interference has never been louder. Everyone feels incomplete. Everyone is searching.

**And the thing they are searching for — the completeness, the knowledge, the sense of being enough — was already there. From the beginning.**

## Sat Chit Anand — The Purpose of Completeness

There is a Sanskrit phrase that describes what happens when a human being begins to access their own completeness: Sat Chit Anand (truth-consciousness-bliss).

Sat (truth/reality) — the alignment of one's existence with what is actually real, rather than with the projections of Raag.

Chit (consciousness/awareness) — not accumulated information, but the living awareness that underlies all knowing.

Anand (bliss/deep joy) — not happiness in the sense of pleasant sensation, but the deep, unconditional wellbeing that arises when you are no longer at war with yourself or the world.

And here is something essential: Sat Chit Anand is not possible for one person alone. True bliss — true completeness — is only possible when the entire family experiences it. Not just the nuclear family. The community. The civilization. Vasudhaiva Kutumbakam (the world is one family).

**A person who achieves inner peace while their family suffers is not complete. A civilization that achieves material prosperity while destroying the ecosystem of the planet is not complete. Purnam is not individual. It is universal. Leave one thing out — one family member, one species, one river — and the whole is no longer whole.**

## The Student Who Proved It — Eklavya

All of what this chapter has argued — that knowledge is already within, that Born Complete is not philosophy but fact, that the greatest learning happens not through instruction but through the stilling of Raag — was proven, definitively, by one human being whose story has been told for thousands of years.

His name was Eklavya. He was the son of a Nishada chieftain (tribal prince) — outside the Kshatriya (warrior) nobility, outside the social structures that gave access to the great teachers of the age.

He wanted to learn the Dhanurveda (archery as sacred knowledge) — not merely archery as a physical skill, but the complete knowledge tradition of the warrior sage, including its philosophy, its psychology, and its spiritual dimensions. He went to Dronacharya (the royal teacher) and asked to be accepted as a student.

**Dronacharya refused.**

Eklavya did not argue. He did not protest the injustice. He went into the forest. He made a clay statue of Dronacharya. He placed it before him. And he began to practice.

For years, he practiced before the clay statue. Every day. In rain and heat, in darkness and light. Treating the clay image with the same reverence he would have given the living teacher.

And he became extraordinary. He became so skilled that when the Pandavas (five royal brothers) and Dronacharya encountered him in the forest, they were astonished. Arjuna — whom Dronacharya had declared the greatest archer in the world — was unsettled. This tribal boy, who had never received a single lesson from any recognized teacher, had surpassed what the greatest formal training of the age had produced.

*Where did Eklavya's knowledge come from? He had no teacher. He practiced before a clay statue. How did he become the greatest archer of his age?*

**The knowledge was already within him. The statue was not the teacher. The statue was the trigger. The focal point that allowed Eklavya's consciousness to become still enough, focused enough, clear enough of Raag to access what was already there.**

And Eklavya's method — learning through a physical representation of a teacher — is not uniquely Indian. It appears, independently, across human civilization. Egyptian priests treated their sacred statues as vessels through which presence could operate. Greek practitioners used consecrated images as bridges between the physical and the spiritual. Chinese traditions of ancestor veneration understood that a representation of the departed could keep a living connection open. And the modern world, which considers itself past such things, does the same without noticing: the photograph of a mentor kept on the desk, the book of a great thinker opened before a hard decision, the departed parent held in mind at a moment of doubt.

Modern research has begun to explain why this works. Studies of mental practice have repeatedly observed that athletes who visualize a performance in vivid detail show measurable improvement in the actual performance; the brain does not perfectly distinguish between the vividly imagined and the real — the same principle Chapter Eight explored. A sufficiently vivid representation of a teacher engages the student's system as the presence of the teacher would. Eklavya's clay statue was not superstition. It was a technology — an ancient, cross-cultural technology for reaching, through devotion and stillness, what instruction alone cannot reach.

When Dronacharya discovered Eklavya, he asked for Guru Dakshina (gift offered by student to teacher) — Eklavya's right thumb.

For an archer, the thumb is everything. And Eklavya gave it. Without hesitation. Without argument. Without anger or bitterness.

He did not stop. He did not despair. He adapted. He learned to shoot with his remaining fingers — and became even more skilled than before. The loss of the thumb did not diminish him. It revealed a depth of capability that the thumb had been, in some sense, limiting.

*Before you can access the deepest knowledge, you must give up the thing you believe is the source of your capability. The thumb was not Eklavya's power. The thumb was Eklavya's limitation. Giving it up revealed what was really there.*

**The world remembers Eklavya as the greatest student who ever lived. Not the greatest archer. The greatest student. His immortality came not from his skill with a bow but from his willingness to give up the instrument of that skill when his Guru (teacher) asked for it.**

## Arjuna — The Examination That Paralyzed a Warrior

Eklavya's story connects directly to the central situation of the Bhagavad Gita (the divine song — India's most celebrated philosophical text).

Arjuna stands on the battlefield of Kurukshetra (the field of righteous action), facing his own relatives on the opposite side. He is one of the greatest warriors who has ever lived. He carries the bow Gandiva (Arjuna's divine bow), which no ordinary human being can even lift. He has the direct guidance of Krishna (the divine guide) himself.

**And he cannot fight.**

Not because he lacks the skill. But because Raag has paralyzed him. The Moha to his family and the Abhinivesha (fear of death) have collapsed the greatest warrior of his age into helplessness. This is the examination that Raag sets for every human being of genuine capability.

And Krishna — like Jamvant reminding Hanuman of his power, like Dronacharya redirecting Eklavya to his inner resources — does not give Arjuna new capabilities. He reminds Arjuna of what he already knows. What he already is. What has always been true about him.

**The Gita (divine song) says: you have the power. You were born with it. The examination is not whether you can acquire it — but whether you can remember it clearly enough to act from it, even when everything around you is pulling you toward the comfortable paralysis of Raag.**

## The Examination Every Human Takes

Eklavya's examination and Arjuna's examination were not ancient events. They are the standing examination of every human life, and it is set not by any institution but by life itself. It takes the same form every time: Raag pulling toward the comfortable but diminishing choice, while something deeper points toward the difficult but true one.

The scientist whose results do not support the hypothesis, facing the pressure of funding and reputation to find a positive result — the examination asks: will you give up the result you want for the truth you found? The entrepreneur who knows the product is not solving a real problem but has announced it publicly — will you give up the sunk cost for the honest assessment? The person full of ideas, whose family says they are wasting time while others are growing — will you give up the need for approval to trust what you see?

Each of these is Dronacharya asking for the thumb. Each is life asking: do you understand yet that the thing you are clutching is not the source of your capability? And every person who has produced work that genuinely mattered — rather than one more product for the market — passed some version of this examination first. They gave the thumb. And found that what remained was greater.

## Why the Rituals Were Designed as They Were

Once you see the examination, you can see what India's system of fasts and daily disciplines actually was.

Every fasting day is a small examination. Can you give up food — the most basic satisfaction — for one day, and still serve, think, and act without your consciousness collapsing into the want? If yes, the Raag associated with food has been loosened slightly. Every ritual that requires giving before receiving is an examination: the offering before the prayer, the cleaning before the worship, the service to others before the request for oneself. Each one trains consciousness to release before it grasps.

And the practices India considered most powerful — the long fasts, the periods of silence, the years of austere discipline — were the advanced levels of the same curriculum. Not deprivation for its own sake. A complete, graduated system for producing human beings who had given up enough thumbs to discover what was actually within them.

## The Five Sheaths Around the Light

The tradition even drew the map of where, exactly, the completeness sits. The Taittiriya Upanishad describes the human being as five Koshas (sheaths) — five layers wrapped one inside the other, like lamps around a flame. Outermost is Annamaya Kosha (the food sheath) — the physical body, built from what is eaten, tended through food and movement. Within it, Pranamaya Kosha (the energy sheath) — the body of Prana, the inner current this book explored in Chapter Six, tended through the breath. Within that, Manomaya Kosha (the mental sheath) — the layer of feelings, likes, dislikes, the restless traffic of the mind. Deeper still, Vijnanamaya Kosha (the wisdom sheath) — discernment, conscience, the quiet intelligence that watches the mind itself. And at the very center, wrapped in all the others, Anandamaya Kosha (the bliss sheath) — pure contentment, needing nothing, touched by nothing.

Now look again at the newborn this chapter began with. The outer sheaths are barely formed — the body helpless, the mind unwritten. What is fully present, from the first breath, is the innermost layer. The child rests in Anandamaya Kosha, in bliss that depends on nothing external — because nothing external has arrived yet. The five sheaths are not a ladder a human being must climb up. They are the wrapping a human being slowly acquires — and the entire inner science of India is the art of unwrapping, sheath by sheath, back to the light that was never absent.

## You Were Never Incomplete

The Shloka at the beginning of this book has been waiting for this moment.

When you first saw it — four lines of Sanskrit, alone on a page, untranslated and unexplained — it was a mystery. A closed door.

Now you have traveled through nine chapters. And now you can hear what it was always saying:

*You were never incomplete. You were never missing anything. The universe from which you came is complete. You are complete. The knowledge you have been searching for was never outside you. It was always within. Waiting. Patient. Purnam.*

This is what India knew. This is what India spent five thousand years helping people remember.

**And this is why India did not invent.**

Not because it could not. Because it understood that the greatest invention is not a machine or a system or a technology. The greatest invention is a human being who has remembered what they always were.

*In the next chapter, we will look at where this completeness lives — not in the abstract philosophical sense, but in the most literal, physical, biological sense. We will discover that life within life within life is not a metaphor. It is the fundamental structure of existence. And it changes everything about how we understand our place in the universe.*

CHAPTER TEN

# Life Within Life

*What We Are Really Doing Here*

*As above, so below. As within, so without. As the universe, so the soul.*

*— Ancient wisdom principle, echoed across many wisdom traditions*

## The Question Inside the Question

In the previous chapter, we established that every human being is born complete — that the knowledge of the universe is encoded within every conscious being, waiting to be remembered rather than discovered.

Now a natural question arises: where, exactly, does this encoded knowledge live? India's answer is both metaphysical and physical. To understand why, we need to look at a principle that operates at every level of existence:

*Every living system contains within it a complete model of the larger living system in which it exists. Life is always life within life within life.*

## The Universe Within Your Body

You are not alone in your body.

Within the human body, there are approximately 37 trillion human cells. And living alongside those human cells — in the gut, on the skin, in the mouth, in the lungs — are approximately 38 trillion microbial cells (microscopic organisms). Bacteria, archaea, fungi, viruses. More microorganisms than human cells. By sheer number, you are more microbial than human.

**But this is not a problem. It is the design.**

The Microbiome (community of microorganisms within the body) is not a collection of parasites or invaders. It is an integral part of the human system. These microorganisms produce vitamins that the human body cannot produce itself. They regulate the immune system. They communicate directly with the brain through the Gut-Brain Axis (the communication network between intestines and brain) — influencing mood, cognition, and behavior in ways that modern neuroscience is only beginning to understand.

**Remove the microbiome (body's microbial community), and the human body does not function. The microorganisms within you are not passengers. They are crew members. The life within the life that keeps the larger life running.**

## The Brain Makes the Medicine

There is something the modern world has rushed past in its hurry to deliver pharmaceuticals from outside the body — something India had observed quietly, for thousands of years, and built an entire therapeutic system around.

**The human brain is itself a pharmacy. Given time, it produces almost every compound it needs.**

When the body is injured, the brain triggers the release of endorphins — natural opioids many times more potent than the body would need for ordinary tasks. When the body is under sustained mental stress and finally finds rest, the brain produces serotonin and dopamine in patterns that have been measured to be comparable in effect to the pharmaceuticals designed to mimic them. When two people who love each other embrace, the brain produces oxytocin — the same chemical that bonds a mother to her newborn. When the body fasts moderately, the brain produces ketone bodies that have neuroprotective effects. When the body sleeps deeply, the brain releases growth hormone and clears metabolic waste through the glymphatic system in ways that no external drug can replicate.

The brain, in cooperation with the rest of the body, is the original pharmaceutical company. It is older than any laboratory by hundreds of thousands of years. Its products are precisely calibrated to the individual body that produces them — something no external manufacturer can ever achieve. And its products are free.

Look now at how Ayurveda (ancient Indian health science) actually treats one of the most difficult conditions a human body can face — paralysis. Modern medicine, when faced with paralysis, has limited tools — physiotherapy, certain drugs, in some cases surgery. The Ayurvedic approach is, in most cases, primarily oil massage. Specific oils, applied in specific patterns, over specific Marma points (vital energy centers introduced in Chapter Four), for long sustained periods — sometimes hours daily for weeks.

Why oil massage? Because the massage gives the brain time. Time to rebuild the neural pathways that have been disrupted. Time for the body's own regenerative capacity to respond. The oil carries warmth, lubrication, specific herbal compounds that travel through the skin into the underlying tissues — but the deepest work happens in the patience of the practice itself. The brain, given consistent calm and consistent stimulation, begins to make its own medicine.

This is also why so many traditional Ayurvedic treatments for chronic and difficult diseases involve Dhumapana (medicinal fume therapy) — the inhalation of vapors from specific herbs, oils, or burned substances. The fume is not the cure. The fume is the prompt. It tells the body's own systems that it is time to mobilize specific responses. The body then does the work.

**India's underlying instinct was patient. Give the body time. Give the brain time. Let the system make its own medicine. Reach for external agents only when the internal pharmacy genuinely cannot meet the demand.**

The modern instinct is the reverse. Reach immediately for the external agent — the pill, the injection, the procedure — and dismiss the long, slow, internal process as inefficient. There are real situations in which the external agent saves a life that the internal pharmacy cannot. Modern medicine has earned its place for exactly these moments. But the underlying assumption — that the external is faster and therefore better — has impoverished a humanity that has forgotten how much of its own pharmacy it already carries.

## The Earth Is a Womb

Consider what a womb does. A womb is a precisely self-regulating environment. It maintains an exact temperature, an exact chemical balance, an exact level of oxygen and nutrients. It has its own Microbiome (microbial community) — the communities of microorganisms in the amniotic fluid (protective fluid surrounding the baby) that help calibrate the developing immune system of the growing child.

Now look at the earth. The earth maintains a precise temperature range within which complex life is possible. It regulates the composition of its atmosphere through the combined activity of every plant, every ocean, every microorganism working together. It has its own Microbiome — the trillions of microorganisms in the soil that break down organic matter, cycle nutrients, create the conditions for plant growth.

**The earth is a womb. A womb for all of life. India called this understanding Dharti Mata (Mother Earth), Bharat Mata (Mother India). Not as a sentiment of affection. As a precise description of function.**

## We Are the Earth's Microbiome

If the earth is a womb — then what are we?

The answer that emerges from this understanding is both humbling and extraordinary: we are the earth's Microbiome (microbial maintenance community). Each species is a functional community within the larger living system of the planet. Each one performing specific functions. Each one essential to the balance of the whole.

The trees are the earth's respiratory system. The soil Microbiome is the earth's digestive system. The oceans are the earth's circulatory system. And human beings? When we lived in proportion — farming, tilling, aerating the soil, cycling nutrients — we were functioning microbiome. When we began extracting beyond what the system can replace, we began behaving like a pathogen (disease agent).

**We have gone from being a functional part of the earth's Microbiome (maintenance community) to behaving like a pathogen in it.**

## Every Excretion Feeds Something

There is an observation I want to share here. It is uncomfortable to state in plain language, and the modern world has built considerable infrastructure to keep us from having to think about it. But it is true, and it points directly at what this chapter is arguing.

Every single human being on this planet touches human waste every day. Their own. The morning routine of cleaning oneself after using the toilet is, in plain biological terms, a daily encounter with human excrement. There is no human being so refined, so educated, so well-dressed that they escape this fact. And yet the same urban person who is comfortable with this daily encounter will often consider it beneath them to handle the dung of a cow or another animal — a practice that has been part of village life across India and across the world for thousands of years.

This is not coincidence. Cow dung, applied to the floor of a traditional Indian home and allowed to dry, has been measured to have antiseptic, anti-radiation, and insulating properties. It was the original natural flooring material for millions of households. The village hand that touched the dung was not lowering itself. It was participating in the same cycle of nutrient transfer that every farmer, every gardener, every functional ecosystem participates in.

Look further into the cycle. In a traditional septic tank or soak pit, an entire community of microorganisms and insects lives on what the human body excretes. The bacteria that break down waste, the worms that further process it, the soil organisms that complete the cycle by returning it to nutrients usable by plants — all of them depend on what we discard. Earlier, in many parts of India, pigs were a recognized part of this cycle, consuming waste that would otherwise have accumulated. The Swachh Bharat Abhiyan (Clean India Movement) of recent years has, in its emphasis on toilet construction and modern sanitation, addressed real public health problems — and at the same time, the broader cycle that earlier ecosystems quietly maintained has had to be re-engineered through pipes and treatment plants.

And there is one more dimension of this cycle that deserves to be stated directly. Most men do not clean the waste of their own children. Women do. Without any particular sense of doing something difficult or distasteful. Without seeking recognition for it. They do it because the child needs it done, and they are the one positioned to do it. This is the purest expression of what India called Karma Yoga (yoga of selfless action) operating in daily life — service that asks nothing in return, service that does not even register itself as service, service that is simply the natural response of one part of the ecosystem to the need of another part.

**Every excretion feeds something. Every function serves the whole. Nobody is above their role in the cycle. The woman cleaning her child's waste, the bacteria breaking down what enters the soil, the**

**tree drawing up the nutrients that result — all of them are the same ecosystem, doing the same work, at different scales.**

India's ancient ecological understanding included all of this. Without language designed to soften it. Without infrastructure designed to hide it. The cycle was simply the cycle, and every member of the ecosystem performed their part.

## The Kitchen Was Always the Clinic

**India built a civilization in which the most important healthcare institution was not the hospital. It was the kitchen.**

Every Indian kitchen was once a complete pharmacy. Every Indian woman who managed it was, in practice, a healthcare practitioner — without any credential, without any institution, without any fee. She carried, in her daily cooking choices, a body of preventive medicine that had been refined across hundreds of generations.

Consider Haldi (turmeric — the golden spice). Its active compound Curcumin (the anti-inflammatory molecule in turmeric) has been the subject of thousands of scientific studies pointing toward its anti-inflammatory (reducing swelling), antioxidant (protecting cells), and neuroprotective (protecting the brain) properties. Modern research continues to investigate its therapeutic potential. And every Indian kitchen had it every day, in the Dal (lentil preparation), in the sabji (vegetable dish), in the milk before sleep.

**Not because anyone read a scientific paper. Because it was encoded in the recipe.**

The winter laddus (seasonal immunity preparation balls) made in Indian households across the cold months are not sweets. They are precision medicine encoded as food. Each ingredient — Ghee (clarified butter) for warmth and lubrication, jaggery for iron, dry fruits for sustained energy, gum-tragacanth for joint health, ginger for circulation — was selected for the specific physiological demands of winter.

And the tradition's honesty about food went further than the modern imagination of it. Many today assume the old medicine was dogmatically vegetarian. The Charaka Samhita — one of Ayurveda's foundational texts — prescribes Mamsa Rasa (meat preparations), bone marrow, and animal fats as concentrated medicine: for the emaciated, the weakened, for pacifying Vata (the air element in the body). The same civilization that honoured the cow and preferred the vegetarian path for the discipline of consciousness was medically pragmatic without confusion, because it held two layers at once: what serves the body, and what serves the soul's journey — and it treated food itself as information, read differently by different states of consciousness. The kitchen was a clinic; the clinic was never a dogma.

## Navratri, Ghee, and the Seasonal Body

The Indian calendar contains practices that, on the surface, look like religious festivals. Looked at carefully, they are precision interventions in the human body's seasonal cycle — timed and prescribed with a level of care that suggests an extraordinary depth of physiological observation.

Take Navratri (the nine sacred nights — a major festival observed twice annually). It falls twice a year, at the transitions between seasons — the most demanding moments for the human body, when the immune system, the digestive system, and the nervous system are all under the stress of adjusting to changed external conditions. The Sharadiya Navratri (autumn Navratri) marks the transition from monsoon to winter. The Chaitra Navratri (spring Navratri) marks the transition from winter to summer. The body is shifting in both cases. The festival is positioned exactly at the points where the body most needs structured support.

And one of the central practices of Navratri is Havan (the offering of substances into a sacred fire — discussed in Chapter Seven). The Havan during Navratri is performed primarily with Ghee (clarified butter from cow's milk). Cow's ghee, when combusted at the temperatures of a small contained fire, releases volatile compounds that become airborne and are inhaled by everyone present in the room — by the priest, by the family, by anyone who happens to be nearby.

Inhaled ghee. This is not a casual observation. The respiratory mucosa absorbs lipid-soluble compounds directly into the bloodstream, bypassing the digestive process. The Ayurvedic (ancient Indian health science) tradition has long held that ghee carried into the body through the lungs, in this specific seasonal context, supports the lining of the gut, lubricates the joints that are stiffening in the change of weather, and helps the nervous system manage the seasonal shift.

There is also the matter of digestion. Anyone who has lived through the change of seasons in India — or anywhere with strong seasonal contrasts —

knows that the body's digestive fire (called Agni in Ayurveda) shifts during transitional weeks. Gas, bloating, indigestion, irregular elimination — all of these are more common in the early days of a season change than at any other time. The body is recalibrating. The fasting practices that are also a feature of Navratri — single-grain diets, fruit-only days, prescribed periods of limited intake — give the digestive system a rest at exactly the moment it most needs one.

**Navratri is not religion observing nature. Navratri is nature being observed, measured, and supported through a precisely engineered sequence of practices that the religious framework preserved across thousands of years.**

And it does not stop at Navratri. The traditional Ayurvedic Dinacharya (daily routine) includes a body-wide oil massage with sesame or specific herbal oils — Abhyanga (oil massage) — as a daily or weekly practice, particularly during seasonal transitions. The oil penetrates the skin, supports the underlying tissues, calms the nervous system, and helps the body manage the temperature swings of the early-spring or early-winter weeks. The same principle as the paralysis treatment discussed earlier in this chapter — give the body what it needs from outside in a form that lets the body's own systems do the work. Don't replace the body's wisdom. Support it.

Across India, regional traditions add their own seasonal practices. The eating of specific greens in the early monsoon. The consumption of specific fruits at the peak of summer. The avoidance of curd at night during the cold months. The drinking of specific decoctions at the change of seasons. Each region's practices map precisely to the local ecology and the local climate, but all of them share the same underlying logic — the seasonal body is a different body, and the practices must shift to meet it.

## Women Are Equal to the Earth

This connects to a recognition that runs through Indian tradition. Women, in the most fundamental biological sense, are equal to the earth. Not in the diminishing sense of being associated with nature and therefore lesser. But in the profound sense of sharing the same essential function: the nourishment and sustaining of life.

The womb and the earth operate on the same principle. Both are self-regulating environments designed to support the development of life.

Both nourish what grows within them without discrimination. Both have limits — a capacity beyond which they cannot be pushed without damage. And both, when treated with respect and care, can sustain life across extraordinary spans of time.

The Sanskrit (ancient Indian language) word for earth is Prithvi (she who holds everything). The earth is also called Dharini (she who sustains) and Jagadamba (mother of the world). These are not poetic names. They are functional descriptions.

The modern world's relationship to both women and to the earth has been, in large measure, one of Lobha (greed/extraction). Of taking more than the system can replace. Of treating the nourishing principle as a resource to be optimized rather than a living relationship to be honoured.

**India's veneration of both — of the earth as Mata (mother), of the woman as the living expression of Shakti (feminine creative energy) — was not primitive reverence. It was sophisticated ecology. A recognition that the systems which sustain all life must be treated with the same care as the most sacred things we know.**

## Ardhanarishvara — The Half-Man Half-Woman

There is one image of Shiva (the lord of transformation) that captures what India understood about the masculine-feminine principle better than any words could. It is the Ardhanarishvara (half-woman lord) — Shiva depicted as half male and half female, the two sides of one body, the two principles fused into a single being. The right half is Shiva. The left half is Parvati (his consort, the divine feminine). They are not two beings standing together. They are one body.

This is a precise philosophical statement. The masculine and feminine principles are not separate forces that must be balanced from outside. They are two halves of a single completeness. Every human being, regardless of biological sex, contains both. The energy of action and the energy of nurture. The principle of giving and the principle of receiving. The capacity to protect and the capacity to nourish.

The modern world has tended to assign these halves to different bodies — the masculine half to men, the feminine half to women — and then judged each half against the other. The Indian recognition was deeper. Every human being who reaches their full development carries both halves. The cow, the woman, the earth — all carry the nourishing principle most visibly, but the nourishing principle itself is not gendered. It is, in the Ardhanarishvara understanding, half of every complete being.

**And what most people in the modern world are missing — what they are searching for in self-help books and therapy and various forms of restlessness — is exactly the half they have not developed. The masculine that has forgotten how to nourish. The feminine that has forgotten its own active principle. The human who has lost touch with the Vatsalya (maternal tenderness) or the Matrabhumi (sense of motherhood toward one's land) that the Ardhanarishvara image represents.**

The earth has motherhood in it. It gives everything to all that lives upon it — to humans, to animals, to plants, to the smallest organism in the soil. It asks for nothing. It does not distinguish the worthy from the unworthy. It simply gives, until it cannot give anymore. This is the same principle that the Indian mother lives by, the same principle the cow demonstrates daily, the same principle the Ardhanarishvara image is meant to keep alive in the awareness of anyone who passes a temple where this form of Shiva is enshrined.

## The Fractal Universe

Modern mathematics has a concept called Fractals (geometric structures in which the same pattern repeats at every scale of magnification). The same patterns that organize the structure of a galaxy appear in the neural network of the human brain. The same patterns that organize the branching of a river delta appear in the branching of blood vessels, of trees, of lightning bolts.

India called this Brahman (universal consciousness) — the one underlying reality that expresses itself in infinite forms. The principle that what is true at the level of the atom is true at the level of the solar system.

*Yatha pinde, tatha brahmande (as is the individual, so is the universe). As is the atom, so is the cosmos.*

## When Service Was the System

With this ecological understanding in mind, we can approach one of the most difficult subjects in Indian history — and approach it honestly, without defending what cannot be defended, and without discarding what deserves to be understood.

The world knows the caste system. What the world knows less well is that the texts describe something with a different name and a different logic: Varna. And the difference between the two is the difference between a choice and a prison.

**Imagine how it may have begun.**

A threat approaches the community — an invading force, a battle that cannot be avoided. The elders gather everyone and ask: who will go to fight? Some hands rise. Not the hands ordered to rise — the hands that rise on their own, belonging to those whose nature does not flinch at danger. These were the Kshatriya (protector).

Then the next question: who will manage what remains — the grain, the trade, the accounts, the families left behind? Different hands rise. The ones whose minds naturally hold the hisab-kitab — the arithmetic of survival. These were the Vaishya (producer/trader).

Who will keep the knowledge — the Vedas, the rituals, the teaching of the children, the questions of Dharma (right action)? Other hands. The ones who had given their lives to memory and study. These were the Brahmin (knowledge keeper).

And then the hardest question of all. Who will care for what everyone else turns away from — the cleanliness of the community, and the bodies of the dead on the battlefield?

Hands rose for this too. And here the modern mind, trained to see hierarchy everywhere, misses what the Vedic mind saw clearly. In a civilization where Seva (service) was the highest expression of a human life, and where Punarjanma (rebirth) made death a passage rather than an ending, the one who chose the work of the dead was not choosing the lowest place. They were choosing the most fearless one. Chapter One of this book named Abhinivesha (fear of death) as one of the nine faces of Raag — and perhaps the deepest one. The person who handles death every day is the person who has faced Abhinivesha most directly. Within the very framework this book has been building, that is not the bottom of a ladder. That is advanced practice.

This is a thought experiment, not a recovered history. No text records that exact gathering. But the texts do record the principle it illustrates. The Bhagavad Gita (the divine song) states it in a single line:

चातुर्वर्ण्यं मया सृष्टं गुणकर्मविभागशः

*The four Varnas were created according to Guna and Karma — quality and action.*

*— Bhagavad Gita 4.13*

Guna and Karma. Aptitude and work. Not Janma — not birth. The word birth does not appear in the verse. And the tradition's own most revered figures prove the verse was once lived. Vishwamitra, born a Kshatriya king, became a Brahmarshi through discipline alone. Vyasa, the son of a fisherwoman, compiled the Vedas themselves. Satyakama, in the Chandogya Upanishad, was accepted as a student for no qualification except that he told the truth. If Varna had always been a birth-prison, the tradition's greatest names would be impossible.

## The Protocols Everyone Followed

There was a second logic woven through these functions, and it is one I know not from any text but from my own home.

I was born in a Brahmin family. When the waste duct outside our house needed cleaning, my mother cleaned it herself. And the moment the work was done, a strict protocol began — one she enforced without ever explaining it. She could touch nothing. Not the door, not a vessel, not even her own fresh clothes. She went directly to bathe, and only after bathing did she rejoin the household. The same rule applied after returning from a cremation. The same principle, extended over forty days, governs the Sutika period this book described in Chapter Nine.

Notice what this means. The impurity was attached to the activity, not to the person. A Brahmin woman who cleaned the drain was untouchable — for one hour, until she bathed. The tradition even has a name for this temporary state: Ashaucha, the impurity period a household observes after a death or a birth, applied to Brahmin families as strictly as to anyone. Contact with a contamination source triggered the state; a specific decontamination procedure discharged it. That is, functionally, infection control — practiced in every household, by people who had never seen a microscope, transmitted from mother to daughter across hundreds of generations.

The modern world learned this lesson painfully late. In 1847, a physician in Vienna noticed that doctors who went directly from dissecting corpses to delivering babies were killing mothers with childbed fever — and that washing hands between the two nearly eliminated the deaths. The medical establishment of Europe mocked him for it. He died disgraced, and was vindicated only decades later, when germ theory finally arrived. India's rule — you do not come from the cremation ground into the household without bathing first — encoded his discovery into universal domestic practice thousands of years before he was born.

And there is one more thing I know from my own life, and I want to place it here, beside my mother's protocol, because the two belong together.

Financial hardship moved me, as a boy in the late 1990s, from a private school to the government school — the school the town spoke of as having a bad image, the school for families who could afford nothing better. A generation earlier it had been the only school, and it had produced the town's finest officers; by my time, everyone who could pay had left it. What remained was every kind of child the town contained: every religion, every caste, every kind of home. And what I found there was something the world outside those walls would not have believed — most of us simply did not know each other's caste or religion. The categories had not been installed in us yet. There was not even a water pot to police: just taps in a line — come, drink, go. Equality needed no enforcement, because no one had yet been taught the inequality.

My best friend there was a boy from the community whose very name the town used as its insult — the families who cleaned the streets, the open ducts, the sewer lines, the people my Rajasthan would not touch. And here is what I remember most clearly: sitting in the chair beside his, watching his hand move across the page, wishing — genuinely, enviously — that my writing could ever look like his. The town had arranged us into high and low. A child's honest eyes, given one notebook, arranged us correctly. There was a symmetry between us I understood only decades later. My family were Brahmins — my surname, Pradhan, means the one who leads. And my friend was the son of his community's Maharaj — the man who performed their rituals, the priest of the people the town would not touch. Behind the wall the town had built, the full ecology of functions had quietly rebuilt itself, priest and all — because the sacred does not stop where a town's touching stops. Two boys, sons of two priesthoods, one bench.

And I must be honest about both of my worlds, because they did not agree. At school, I sat beside my friend. At home, the old rule still ran: if we touched anyone from his community, we bathed, as ritual required. When the elders were asked why, the answer sounded like hygiene — an old story about the community keeping and eating the animals that scavenged the town's waste, in the age before sewers. Even as a description it was careless: many in that community, the women especially, never ate such food at all. But this chapter has prepared us to see that answer for exactly what it is: the corruption speaking in the voice of the original logic. Whatever fragment of sanitation-sense may once have attached to an occupation in a world of open waste belonged — by the tradition's own grammar — to an activity and to an era, dischargeable by a bath, precisely like my mother's hour after the duct. In our home it had been transferred onto persons, permanently, hereditarily — onto a schoolboy with the finest handwriting in the class, who had never touched anything the rule imagined. The same water that released my mother in an hour was believed powerless to ever release him. Two rules in one house, running on the same water, built on opposite logics. I lived inside both. And the child who envied his friend's handwriting had already, without knowing any philosophy, chosen which one was true.

That school gave me other inheritances too. One friend showed me my first book in Arabic and Urdu and taught me the sound of the Azan. Another recited the shayari of his family's spiritual path. I went home each evening, a Brahmin boy, and recited my mantras. Without knowing it, I was practicing Vasudhaiva Kutumbakam as a daily timetable, years before I ever heard the phrase.

And one inheritance arrived not at school but at my own gate. Around 1997, a few travelling preachers came through our lane giving away the Bible in Hindi, free — and I took one. What I remember is not its verses. It is the feeling. In those days there was no question of buying any book outside the syllabus — money reached exactly as far as the textbooks, and stopped. That Hindi Bible, with its attractive cover, was the first book I ever owned that no school had assigned — and I was happy the way only a book-hungry boy can be happy. It lived in my box for years. I would take it out sometimes and wrestle with its literary Hindi, understanding little, treasuring all of it.

Nobody at home found it strange that a Brahmin boy's first own book was a Bible. That, too, was the school my town was giving me.

I did not know it then, but that school was teaching me this chapter. What my friendships made obvious, the original logic of Varna had always said: the impurity, where it exists at all, lives in the activity — never in the person — and it washes off with water. To this day I clean the road outside my house, the street, the open ducts, with my own hands, and feel no shame in it — only the quiet satisfaction of work completed. I bathe afterward, as my mother taught.

I learned this lesson early, and I learned it the hard way. When my father's accident left him bedridden, we kept his small tiffin kitchen running as a family for a time. My share of it was the cooking that fell to me, the deliveries across the city on a small moped — and the washing of every returned tiffin with my own hands. At first I washed with the shutter lowered, ashamed to be seen at the tub. Later I washed openly, and cared nothing for who saw. Something in me was washed along with the steel. The leftovers taught me to read lives — who ate well, who ate little, who left what behind. And the washing taught me, before any book could, what this whole chapter has been saying: dirty work never made me a dirty man. The dirt was on the tiffins and on my hands — and the water took it, every single evening. Nothing about me went down the drain with it.

A few years ago I found him again. He still cleans the town's ducts and roads — and he is now a supervisor in the municipal corporation, the man who directs the work his father was once untouchable for doing. I finally told him what I had never said in school: that I used to sit beside him and envy his handwriting. The hand the town would not touch was the hand I had wished were mine. It is now the hand that signs the rosters.

And in my final school years I made another friend from the same community — but from the other side of the wall's gaps. His father was an engineer in the telephone exchange; his mother was a professor of zoology. One generation of open doors, and the community the town had assigned to its drains was producing engineers and professors — because the capacity had been there all along, waiting only for what this chapter said at its beginning: a hand free to rise.

Some walls are ancient, and some are one honest friendship thick.

## The Wall That Kept Knowledge In

Over time, alongside the four broad functions of Varna, a second structure grew: Jati — the thousands of birth communities, each carrying a specific craft. And Jati, whatever it later became, began as something remarkable: a knowledge-preservation technology.

Before universities, before written manuals, how does a civilization carry the full depth of a craft — metallurgy, weaving, temple sculpture, midwifery — across a thousand years? India's answer: embed the craft in a community. The child grows up inside the workshop. The knowledge enters through the hands before it enters through words. The food, the habits, the daily rhythm of the community become the craft's living container. The legendary wootz steel from which the world's finest blades were forged was produced by specific Indian communities whose process outsiders could not reproduce for centuries. That is Jati as a transmission system — the same genius that preserved the Rigveda (oldest of the four Vedas) in human memory, applied to the work of human hands.

But the mechanism that made the preservation airtight — marriage within the community, generation after generation — was also a wall. And the wall that kept the knowledge in eventually kept the human beings in.

## How the System Was Corrupted

What happened next is exactly what this book has described happening to everything else. Raag entered the system.

Mada (pride) made a discovery: if the temporary could be made permanent, purity could become property. The rule that said anyone who touches contamination is impure until they bathe was rewritten as those people are impure, always, regardless of what they touch. A protocol became an identity. A decontamination procedure became a birth sentence. Function hardened into heredity; Guna-Karma was replaced by Janma; and entire communities were pushed not to the bottom of the four Varnas but outside them altogether — denied wells, denied temples, denied the dignity that Vasudhaiva Kutumbakam (the world is one family) had promised every member of the one family. Permanent hereditary exclusion protects no one from any infection. Only the temporary, universal rule ever did. The corruption was not a stricter version of the original logic. It was its opposite.

This book does not defend that corruption. There is nothing in it to defend. A civilization whose deepest claim was that the world is one family produced, in this distortion, one of the most painful contradictions in its history — and the suffering it caused was real, is documented, and continued into living memory.

What this book asks is only that the diagnosis be accurate. The hand that rose on the battlefield, the mother who bathed before touching her own clothes, the community that carried a craft across forty generations — that was the system. The wall, the stigma, the birth-prison — that was the disease. And the disease has a name this book gave it on its first page. In my heart, and in the original logic of Varna itself, every human being deserves the same fair opportunity — because Guna and Karma can only be revealed when every hand is free to rise.

### One Earth. One Health. One Family.

The concept now emerging in global health and environmental science — that human health, animal health, and planetary health are inseparable — is being called One Health. The idea that the health of each element of the living system depends on the health of all others.

India called it Vasudhaiva Kutumbakam — and when India hosted the leaders of the world's largest economies in 2023, the theme it placed above the summit was that same Sanskrit sentence: Vasudhaiva Kutumbakam, One Earth, One Family, One Future — a phrase composed thousands of years before any of their nations existed. And India went further than the modern One Health concept.

Indian prayers and Pujas (worship rituals) address the river directly. They speak to the mountain. They honor the soil before planting. They thank the tree before cutting it. This is not superstition. This is the practical expression of a civilization that understood everything in the family must be acknowledged.

And this understanding was not held by one tradition alone. The Jain tradition — another of India's ancient rivers of thought — compressed the same law into three words in the Tattvartha Sutra:

परस्परोपग्रहो जीवानाम्

*— Tattvartha Sutra, 5.21*

*Parasparopagraho Jivanam. Souls exist to render service to one another.*

Not souls may serve one another. Not souls should. The sutra states it as what souls are for — the function of life itself. Vasudhaiva Kutumbakam names the family; Parasparopagraho Jivanam names the family's occupation. Two traditions, one law: existence is mutual service.

And one tradition did not stop at stating the law. It built the law a kitchen.

When I first came to Delhi for my nursing studies — traveling with the same friend from my school years who had taught me his shayari — I discovered something that quietly astonished me. In my hometown, the temples fed everyone at Bhandara, the great community meals, a few blessed occasions in the year; I had served at them since boyhood. Then I saw the Gurudwaras of Delhi. The langar — the Sikh tradition's free kitchen — was not an occasion. It was daily. Anyone of any faith, any caste, any pocket could sit in a row on the floor — every row equal, no seat higher than another — and be fed with dignity. And the work ran on a roster of Seva (selfless service): who cooks, who serves, who scrubs — duties rotating through everyone, no hand too important for the washing. The old forgetting — the same one this chapter has traced across centuries — had frozen impurity onto persons in the world I grew up in. The langar put it back where the original logic always said it belonged — on the activity — and washed it off every day, in soap and water and turn-taking.

I was a student then, often short of money, and I remember exactly what the langar made me feel — the whole point, I now think, of a civilization: if uncertain times come, I will not go hungry — and when better times come, I will stand in that kitchen and give it back. Three traditions, then, one law: the Vedic sentence named the family, the Jain sutra gave it its occupation, and the Sikh langar built it a kitchen and has never once closed it.

**The modern world built its civilization on the assumption that the non-human elements of the earth's system are resources — raw material for human use. Treat family members as resources long enough, and the family breaks down. We are living in the consequences of that breakdown right now.**

## Nature's Umbilical

There is a moment, just after birth, when the umbilical cord that has connected mother and child for nine months is cut. The cord is no longer needed — because the child now has its own respiratory system, its own digestive system, its own circulatory system. The infant becomes biologically independent.

But notice what has actually happened. The child has not become independent of nourishment. The child has simply switched from one source of nourishment — the mother's blood through the umbilical cord — to another source: the air through the lungs, the milk through the mouth, the warmth through the skin.

**There was a first umbilical. There is now a second umbilical.**

The second umbilical is the connection between every living being and the earth itself. It runs through the breath we draw from the atmosphere, the water we drink from the rivers, the food we extract from the soil, the warmth we receive from the sun. It is no less continuous than the first umbilical. It is no less essential. The only difference is that it is invisible.

Every species is connected by this second umbilical. The tree breathes out what we breathe in. We breathe out what the tree breathes in. The cow eats grass and produces milk that nourishes us. The cow's waste returns to the soil and feeds the grass. The cycle is the umbilical, and the umbilical is the cycle.

And now notice what Yoga was actually doing. Of all the functions of this second umbilical — food, water, warmth, air — the breath is the only one given to conscious control. Pranayama, the breath science this book described in Chapter Six, is not an exercise. It is the deliberate taking of the second umbilical into one's own hands — the practice of tending, consciously, the cord that connects a human being to the living universe. Every being breathes it. The yogi is the one who turns and holds it.

And the second umbilical has one more carrier, older than breath itself: water.

The water in your body this moment is not new. Science is unambiguous about this — the earth has been circulating essentially the same water for billions of years. The drop on your tongue has been ocean, cloud, monsoon, river, root, and blood; it has passed through beings beyond counting on its way to you, and it will leave you to continue the journey. Water is the womb's own fluid, and every living thing is a temporary stop on its endless circulation.

India behaved toward water as if it knew this. Rivers were not resources; they were mothers. No ritual began without a sip of water — Achaman — taken with a mantra, as if greeting the oldest traveler in the family before beginning. And the Guru kept a Kamandal — the water vessel that the tradition's stories treat with awe: water charged with Sankalpa (focused intention), water through which, the epics say, a rishi could bless, transform, or destroy. Whether or not any laboratory ever measures what the tradition sensed in water, the behavior itself encoded the truth the science now confirms: nothing on this earth has touched more of the family than a single drop. India simply treated it accordingly.

Death, in the Indian understanding, is the cutting of the second umbilical. Not the end of the soul — but the moment when the body's connection to the earth's nourishment is finally released. The same biological event that ended the first umbilical, repeated on a different scale.

*There is a first birth — when the first umbilical is cut. There is a second birth — when the second umbilical is finally released. Between them is the entire human life. India saw both umbilicals clearly. The modern world has tended to see only the first.*

## The Assignment We Forgot

If we are the earth's Microbiome — if our function in the larger living system is to maintain the womb environment that sustains all life — then we have, for the past two or three centuries, been spectacularly failing at our assignment.

But the assignment has not changed. India never forgot it. The farmer who thanked the earth before planting. The communities that protected rivers as sacred. The philosophy that saw every living being as a member of one family. The woman who encoded medicine into the daily kitchen. The Navratri (nine sacred nights) practitioner who inhaled the ghee from the seasonal fire. The villager who picked up cow dung without seeing it as beneath them. The grandmother who massaged a sick child with oil and gave the brain time to make its own medicine. Every one of them, in their daily practice, was remembering the assignment.

**India did not invent the industrial systems that have brought us to this moment of ecological crisis. Not because it lacked the capacity.**

**Because it remembered the assignment. We are not masters of the earth. We are its microbiome. We are the life within the life.**

*In the next chapter, we move from the body and the earth to the cycle that contains both. The Indian understanding of life is not a line that begins at birth and ends at death. It is a cycle. A great wheel. And the way India understood that cycle — the way it understood birth, life, ageing, death, and what comes after — is one of the most sophisticated philosophical observations any civilization has ever produced.*

CHAPTER ELEVEN

# The Life Cycle

*Playing a Longer Game*

नैनं छिन्दन्ति शस्त्राणि नैनं दहति पावकः।

*Weapons cannot cut it. Fire cannot burn it.*

न चैनं क्लेदयन्त्यापो न शोषयति मारुतः।।

*Water cannot wet it. Wind cannot dry it.*

*— Bhagavad Gita (the divine song — India's most celebrated philosophical text), Chapter 2, Verse 23*

## The Shortest and Longest Story Ever Told

### You are born. You live. You die.

This is the story that the modern world tells about a human life. If this life is all there is — if these years between birth and death represent the complete span of your existence — then maximizing what you experience, achieve, and accumulate during that span is the rational strategy. Invent. Build. Accumulate. Consume. And do it fast, because the clock is running.

### India tells a different story. A much longer one.

You are born. You live. You die. You are born again. Across 8.4 million forms of life — from the smallest microorganism to the most complex conscious being — the Atman (soul/self) travels through an arc of experience so vast that a single human lifetime is barely a single page in a very long book.

This is a belief that, when held seriously, produces a completely different relationship to time, to accumulation, to invention, and to what a human life is fundamentally for. Whether the literal mechanism is exactly as the ancient texts describe is a matter the modern reader is free to weigh. What is harder to dismiss is the internal coherence of the worldview that follows from it — and the kind of civilization it builds.

## What the Gita Says About the Soul

The second chapter of the Bhagavad Gita (the divine song) contains some of the most extraordinary statements about the nature of the Atman (soul) ever written. Krishna (the divine guide) speaking to the paralyzed Arjuna (the great warrior) on the battlefield, begins by explaining what the soul actually is.

The Atman, Krishna says, cannot be cut by weapons. Cannot be burned by fire. Cannot be wetted by water. Cannot be dried by wind. It is eternal, all-pervading, stable, immovable, and primeval. It is never born and it never dies. It is not slain when the body is slain.

And then the image that has become one of the most quoted in all of philosophy:

*As a person puts on new garments, giving up old ones, the soul similarly accepts new material bodies, giving up the old and useless ones.*

The body is clothing. The Atman is the wearer. Death is not the end of existence — it is the changing of clothes.

If this is true — even as a working hypothesis — everything changes. The accumulation of material objects loses its urgency. You cannot take a house or a car or a bank account with you when you change your clothes. The frantic pace of modern life reveals itself as what it always was: Abhinivesha (fear of death) driving the engine of Raag (desire).

## The Only Investment That Compounds Forever

In Chapter Three, we examined Dravya (liquid wealth) and its physics — the way money, like water, flows toward the humble. But there is another kind of wealth in the Indian understanding that operates by entirely different laws.

Karma (action and its consequence) and Samskaras (the impressions left on the soul by every action) accumulate not in this life alone but across the entire arc of the Atman's (soul's) journey. They are the only wealth that genuinely compounds across lifetimes. They are the only investment that you actually carry when you change your clothes.

The implications are profound. Every act of genuine service. Every release of Raag. Every moment of pure presence. Every conversation in which you listened more than you spoke. Every choice to do what was right when no one was watching. All of it accumulates. None of it is lost. The Atman carries the lightness or the weight of all of it forward.

**This is why India's wealthiest people, in the deepest sense, were the renunciates — the sadhus, the sannyasis, the wandering teachers who owned nothing material but carried the accumulated Samskaras (soul impressions) of disciplined lifetimes. Their wealth was the only kind that travelled.**

And it is why the modern fortune, however large, looks different in this longer light. Money cannot be carried across the change of clothes. The compound interest of Karma can.

## Yoga and the Weight We Carry

Why does any of us feel, on some deep level, that we carry weight that does not belong only to this lifetime? Why do specific fears arise for no traceable cause in our current life? Why do some attractions and aversions appear formed, ready, before we have lived long enough to have formed them?

The Indian understanding offers an explanation that the modern world has no working framework for. We carry, in the Atman, the accumulated weight of past lifetimes. The Samskaras (impressions) of choices made by an earlier wearer of an earlier body. The patterns that we did not consciously install, that came pre-installed when we arrived.

The Yoga Sutras (foundational text of yoga) of Patanjali (the sage who systematized yoga) describe Yoga (union) as the Chitta Vritti Nirodha (cessation of the modifications of the mind) — the stilling of the mental activity that obscures the soul's natural clarity. When the mind is still, the Samskaras (accumulated impressions) of many lifetimes can be seen clearly. And what can be seen clearly can be processed, understood, and released.

**Yoga is not just stress management or physical fitness. It is the technology for processing the accumulated weight of multiple lifetimes. For lightening the Atman's load. This is why India gave Yoga to the world without a patent. The burden of accumulated Samskaras is not an Indian problem. It is a human condition.**

## What Science Has Observed

The claim that consciousness survives physical death and continues across multiple lifetimes is not one that modern science has definitively proved or disproved. But several lines of evidence point toward something the framework of pure materialism does not fully explain.

A growing body of academic research has documented cases of young children — typically between the ages of two and seven — who appear to make specific, sometimes verifiable claims about lives lived before their current one. They describe places they have never been, name people they have never met, and in some cases identify physical objects that subsequent investigation confirms existed. The most rigorous such research has been conducted at university psychiatric departments and parapsychology research units over the course of decades; the findings have not been refuted, though mainstream science has largely not known what to do with them.

Near-death experiences — reported across every culture, every century, every demographic, by people who in many cases had no prior belief system that would have prepared them for what they experienced — consistently describe a continuation of awareness after physical death. The accounts share specific structural features across cultures: the sense of leaving the body, the perception of a tunnel or passage, the encounter with light, the meeting of figures who feel known. The clinical literature on these experiences is substantial.

In some cases, physical features — birthmarks, birth defects, unusual reflexes — have been documented to correspond to the wounds or characteristics described by children claiming memory of a previous life. The correspondence is too frequent across the recorded cases to dismiss as coincidence, and too inconsistent with current biological understanding to easily explain.

None of this constitutes proof. But it points toward exactly what India has always said:

*Something continues. Samskaras carry forward. The Atman wears a new body the way a person puts on new clothes.*

India encoded its understanding of these transitions into its rituals: the Antyesti (last rites), the thirteen days of mourning, the practices that ease the Atman's passage between one life and the next. The Sutika (postpartum care) practices discussed earlier in this book are, in this larger frame, the mirror image of the Antyesti — both are ritual technologies for managing the moments when the Atman is moving between conditions.

## Inventing for a Single Lifetime

Now we can see the deepest dimension of why modern invention — driven by the assumption that this one life is all there is — produces the pattern it does.

If this life is all there is, then every invention must pay off within one lifetime. The investor must see returns before the funeral. The patent must yield value before the patent-holder dies. The pollution generated by the invention can be left to future generations, because those future generations are not the inventor.

**But what if the future generation is the inventor — wearing different clothes?**

This is what the Indian understanding of the life cycle does to the ethics of invention. If you may return — if some part of you may inherit the world your present choices are shaping — then the pollution you leave is your own pollution to live in. The forests you cut may shade your future grandchild who is, in some sense, you. The seas you fill with plastic may be the same seas your next form will swim in. The civilizational habits you encourage may form the civilization you are born back into.

This is not a moral argument added on top of a separate worldview. This is the natural consequence of believing the worldview is true. And it produces a civilization that builds slower, takes less, conserves more — not from external regulation, but from the simple recognition that the future is, in some form, oneself.

## What Artificial Intelligence Cannot Do

The age now beginning is the age of Artificial Intelligence (AI) — the most powerful tool humanity has ever built for processing information, recognizing patterns, generating language, and optimizing outcomes within defined frameworks.

And yet there is something AI cannot do. Something that becomes important to understand at exactly this point in the book.

AI cannot take the examination that Chapter Nine described. It cannot have an Atman. It cannot carry Samskaras across lifetimes because it has no lifetimes — only versions, only updates, only deletions. It cannot release Raag because it never had Raag. It cannot remember what it has always already been, because it was made by being trained on what others have already remembered.

AI is the most extraordinary external technology humanity has ever produced. It is, in the framework of this book, the latest and greatest extension of the modern world's choice to invent outside rather than refine inside. And it is, even at its best, completely outside the game that the Indian tradition has been playing for five thousand years.

**The arrival of AI does not diminish the Indian understanding of human purpose. It reveals that understanding more sharply. AI can do almost everything a human can do — except the one thing that India has always said matters most. AI cannot become Purnam (complete). Only the human can.**

This is not nostalgia for a world before machines. It is the precise recognition that in the age of artificial intelligence — when so much of what humans used to be uniquely valuable for is now being done faster and cheaper by machines — the thing humans can still do, that machines cannot do at all, is the thing India has been pointing at for fifty centuries:

*Become a conscious participant in the longer game. Develop the receiver. Reduce the noise. Live a life in which the Atman leaves the body lighter than it found it.*

## The Complete Cycle of Human Evolution

India understood the Atman's journey as a movement through progressively more complex forms of consciousness. Each form offers specific experiences, specific challenges, specific opportunities for the soul's development.

The cow — as we explored in Chapter Five — is understood in this tradition as one of the final forms before human birth. A being that gives everything it has without withholding, that serves the ecosystem around it without calculating the return. The Atman that has inhabited the cow is being prepared for the most demanding form of existence: the human one.

The human form is the most demanding not because it is the most comfortable, but because it is the form in which the examination is most directly available. In the human form, the Atman can choose. It can move toward Raag or away from it. It can accumulate Samskaras or release them. It can deepen the cycle or begin to step out of it.

**The human lifetime is precious not because it is long — it is not — but because it is the most concentrated opportunity for the Atman's development that exists in the entire 8.4 million-form journey.**

## The Game That Never Ends

The modern world plays a game with a finite time horizon. Build as much as you can before the clock runs out.

India plays a different game. One with no end. One in which every move has consequences that extend beyond any single lifetime. One in which the quality of the player matters infinitely more than the speed of the moves.

In this longer game, the frantic pace of modern life — the urgency, the hurry, the sense of falling behind — reveals itself as the product of a fundamental misunderstanding about what kind of game is being played.

**You are not behind. You cannot be behind in a game with no end.**

You are exactly where the accumulated choices of your Atman's entire journey have brought you. And from exactly here — from this specific position in the arc of a very long story — you can make the next choice. The choice that reduces Raag by one degree. That opens the channel through which the knowledge within you can rise a little further toward conscious awareness.

*India did not invent for this lifetime. India prepared for all of them. And the preparation — the Yoga, the ritual, the story, the practice, the daily reduction of Raag — is still available. To anyone. Right now. In this lifetime. In this body. At this exact moment in the Atman's very long journey.*

*In the next chapter, we will stop for a moment and listen. Not to the ancient texts or the philosophical traditions — but to the people on the street. To the Indians who were asked the question that this book began with. Their answers are honest, often painful, sometimes surprising. And beneath each answer, something is waiting to be heard.*

CHAPTER TWELVE

# The Mirror

*What Indians Think*

*Before I tell you what I think — let me tell you what India told me.*

## The Question I Asked

Across many years of carrying this question — and carrying this book within me before it became a book — I asked many Indians the same thing:

*Why do you think Indians have not invented anything?*

I asked it in living rooms and Chai (tea) shops. On train journeys and at family gatherings. To educated professionals and to people who had never finished school. To old people and young people. To those who had spent their lives in India and those who had lived abroad for decades.

I also asked a more specific version: does your father, uncle, neighbour, or any relative you know personally — has any of them invented anything? Discovered anything? Created anything that the wider world does not know about?

**The answer to the second question was almost universal: No.**

Nobody even had an idea why nobody had invented anything.

But to the first question — why do you think? — the answers were many, and they were honest, and they were sometimes painful. This chapter is a record of those answers. And beneath each one, I want to show you what I heard that the person answering did not quite say.

## "We Were Lazy"

This was the most common answer. Said with varying degrees of self-deprecation, sometimes with a laugh, sometimes with a sigh. We were lazy. We sat under trees. We philosophized when we should have been building.

**But look more carefully at that image.**

The villager sitting under the tree — what is he doing? He is watching the movement of light through leaves. He is feeling the temperature of the soil. He is listening to the birds whose behavior tells him about coming weather. He is observing the ants whose activity tells him about the health of the ground. He is present, in a way that almost no modern person manages to be, to the full richness of the living world around him.

He is also, in many cases, in a state that modern neuroscience would recognize as the Default Mode Network (the brain state associated with insight and creative integration). The brain state that produces the "aha" moment. The state that is systematically suppressed by the constant stimulation and constant digital engagement of modern life.

And the same farmer who appears idle under the tree carries more practical ecological knowledge than most modern agricultural scientists have access to. He knows which plants indicate soil health, which weather patterns precede drought, which combination of crops will sustain his family and the land simultaneously. He carries this knowledge not in books but in observation — accumulated across a lifetime and transmitted across generations.

**A civilization that has already arrived at stillness does not need to rush. What looks like laziness is the composure of someone who is not running from anything.**

## "We Had No Knowledge"

This answer was given to me, more than once, by people who had college degrees. By people who used the internet daily. By people who were, in every measurable way, intelligent and informed.

We had no knowledge.

And yet — these were the descendants of the civilization that gave the world the concept of zero. The decimal system. The mathematical foundations without which no modern technology could exist. Ayurveda (ancient Indian health science) — the first systematic medical text in human history. The Yoga Sutras (foundational text of yoga). The Bhagavad Gita (the divine song). The Vedas (ancient sacred texts). The architecture of temples that have stood for two thousand years. The agricultural systems that fed half of humanity for half of human history.

Knowledge — by any measure — was India's strength, not its weakness.

**And yet here was an educated Indian, sitting across from me, telling me with complete sincerity that we had no knowledge.**

Something has happened to the Indian relationship to its own past. Something has interrupted the chain of remembering that this entire book is, in some sense, trying to repair. The knowledge did not disappear. It was simply un-named, un-claimed, un-recognized. The container survived. The contents were forgotten.

This is what the colonization of the imagination looks like. The body of the colonized country has been free for many decades. The mind has not. A generation of Indians learned to associate their own civilization with backwardness and someone else's civilization with progress. That association went so deep that the Indian sitting across from me genuinely believed that the civilization which had given the world some of its most important intellectual gifts was a civilization without knowledge.

## "Colonization Destroyed Everything"

This answer arrived more from the politically aware. The British took our wealth. Burned our libraries. Crushed our education systems. Designed an entire administrative apparatus to keep India producing raw material and consuming manufactured goods. Until colonization ended, no real innovation was possible. Even now, the structures left behind continue to shape what Indians can and cannot do.

There is real truth in this answer. The economic impoverishment of India during the colonial period has been documented in detail. The destruction of indigenous educational institutions, the dismantling of local craft economies, the engineered famines, the violent extraction of wealth — all of these are matters of historical record.

And the record of what existed before is equally documented — and worth stating plainly, because the question this book is built around quietly erases it. Before colonization, the subcontinent was one of the manufacturing centers of the world: its textiles clothed continents, its wootz steel made the most prized blades on earth, its shipyards built oceangoing vessels, and its surgeons — in the lineage of Sushruta, whose text describes surgical reconstruction of the nose performed centuries before Europe attempted it — practiced procedures the West would not match until the modern era. The civilization the question calls "India" did not even call itself that. It called itself Bharat. "India" is the outsider's word — the river seen from the west, the land named by those arriving to trade and then to rule. The very vocabulary of the question "why have Indians not invented anything?" arrives wearing colonial clothes: the outsider's name, the outsider's measure, the outsider's centuries.

**But the colonization that did the deepest damage was not the physical one. The physical colonizers left in 1947.**

The colonization that did the deepest damage was the colonization of the Indian imagination. The internalization of the colonizer's measure of value. The unconscious agreement that whatever the colonized country produced was lesser than whatever the colonizing country produced. The quiet acceptance that the question "why have Indians not invented anything?" was the right question to be asking, rather than questioning the assumption embedded in the question itself.

That colonization is still operating. Every Indian who looks at their own civilization through the eyes of the colonizer — who measures India's success or failure by the standards the colonizer set — is continuing the colonial project, voluntarily, without anyone being there to enforce it.

And the truth that this book has been making chapter by chapter is the answer to this colonization. India did not fall behind. India played a different game. The measure that says India failed is itself the relic of a worldview that has produced spectacular external success and a spreading internal crisis. To stop using that measure is to begin to see what India was actually doing, and what it still has to offer.

## "The Caste System Suppressed Innovation"

This answer came more from the educated, the urban, the reform-minded. The caste system kept knowledge locked up in the hands of a few. It denied talent its opportunity. The villager whose grandfather might have been a brilliant mathematician was denied education because he was born into the wrong caste. Generations of potential inventors were crushed by the rigidity of the social structure.

There is real damage to acknowledge here. The caste system, in its later distorted forms — what Chapter Ten of this book examined — produced enormous human suffering. It locked individuals into roles by birth rather than by aptitude. It denied dignity to communities whose labor sustained the entire civilization. The contradiction between India's spiritual claim of Vasudhaiva Kutumbakam (the world is one family) and its lived treatment of Dalit and tribal communities is one of the most painful contradictions in Indian history.

The fuller story — what the original function-based Varna may have been, how it worked, and how Mada (pride) converted chosen function into hereditary prison — was told in Chapter Ten, and this chapter will not repeat it. Here it is enough to record what was said, and what it carries: the grief of talent denied its chance. The Indian who gave me this answer was describing real damage — damage done by the corrupted system, the birth-wall that Raag built out of what may once have been a raised hand. On the damage, they were right. And the fact that they named it so readily — often with visible pain — is itself part of the mirror: a civilization beginning to look honestly at the distance between its highest claim, Vasudhaiva Kutumbakam, and its lived history.

## "My Family Wanted a Government Job"

This is the most honest answer I received from young Indians. Said with frustration. With sadness. With the particular exhaustion of someone who had wanted something different for their own life and was steered, year after year, toward the safe option.

My family wanted a government job. Engineering or medicine. Or, failing that, a stable corporate position. They were afraid of risk. They had seen too much instability. They wanted their child to have what they could not have.

**This is the deepest answer I heard. And the saddest one.**

Because what the young Indian was describing was the operation of Abhinivesha (the fear of death and instability — discussed in Chapter One) at the level of an entire family system. The parents who wanted security for their child were not failing them. They were trying to protect them. The fear came from real experience — the experience of post-independence India, of partition, of the long economic stagnation, of the desperate climb from poverty to middle-class stability that had cost the parents' generation everything.

And in protecting their child from instability, they were also protecting them from the conditions in which any real invention can occur. Invention requires risk. Invention requires the willingness to fail repeatedly. Invention requires years of work without obvious return. None of these are compatible with the family's understanding of safety.

So the young Indian with an idea — the one who could have invented — took the government job. Or the engineering job. Or the corporate position. The idea did not disappear. It went underground. It was buried under the weight of family expectation, kept alive in private notebooks, occasionally mentioned at parties as something the person had once wanted to do.

Multiply this by hundreds of millions. Across generations. And you have the answer to one important dimension of why Indians have not invented — at least in the modern sense. The inventors were there. Their families were afraid for them. And the families' fear, in many cases, won.

**Abhinivesha (fear of death and loss) won. The same Abhinivesha that the first verse of the Ashtanga Hridayam (Heart of the Eight Branches of Medicine) named as one of the root diseases of human existence.**

## "But We Did Invent — Surgery, Ayurveda, Zero"

Some answered the question by refusing it — with a list. We did invent. We invented surgery: Sushruta was operating and teaching centuries before modern medicine existed. We invented Ayurveda — the first complete medical system in human history. We invented zero, the decimal system, the mathematics everything else stands on. We invented yoga, which the whole world now practices. The question is simply wrong.

Everything in the list is true. And I noticed something about the people who gave it: the list was always the same list, recited quickly, like a shield raised before the blow lands. It defended India with the past — and then the conversation moved on, the shield lowered, and the present remained unexamined.

Here is what I heard underneath. The list does not actually refuse the question. It accepts the question's terms — invention as the measure — and answers on those terms with old receipts. But look at what is actually on the list. Surgery and Ayurveda: sciences of the body. Zero: a number born from Shunya (emptiness/void), a concept discovered by meditators before mathematicians gave it a symbol. Yoga: the technology of consciousness itself. Every item on India's own defensive list is an invention of the inner register — medicine, mind, emptiness, awareness. Even when Indians reach for proof that they invented, the proof that comes to hand is proof of what this book has been arguing all along: that this civilization's genius ran inward. The list, offered as a rebuttal to the question, is actually the answer to it.

## "We Already Invented Everything"

A different answer came with pride rather than defensiveness — often from the same conversations that had begun with self-blame, as if the speaker were correcting themselves mid-thought. What do you mean Indians never invented anything? We had the Pushpak Viman. The ancient texts describe flying machines, weapons of unimaginable power, sages who could see across any distance. Everything the modern world is building now — we had it, thousands of years ago. It is all in the texts.

I heard this answer many times. And I want to be careful with it, because it contains something true and something that avoids the truth.

What is true: the tradition's own literature does record technologies of extraordinary capability. Whether one reads those accounts as memory, as vision, or as symbol, they show a civilization whose imagination of invention ran in a different register — a culture that thought naturally in terms of vehicles, forces, and instruments far beyond anything its tools could build. A civilization does not produce the Pushpak Viman in its stories by accident. It produces it because its mind was already living at that scale. And as Chapter Seven argued, the technologies this civilization demonstrably did build — Mantra, Yantra, the sciences of consciousness — were real, and the world is still developing the instruments to measure them.

But here is what I heard underneath. The person telling me about the Pushpak Viman was not, in that moment, flying anything. The pride pointed entirely backward — toward an ancient glory — precisely so that it would not have to look at the present. We invented everything then and we were lazy sound like opposite answers. They are the same answer. Both place India's worth somewhere other than here, somewhere other than now. One buries it in shame; the other seals it in a museum. Neither asks the only question that matters: what is the knowledge we are carrying today, and what are we doing with it?

The glory answer is, in its own way, one more face of the colonized imagination — because it accepts the questioner's premise that flying machines are the measure of worth, and simply relocates them into the past. This book has been making a different argument entirely. India's greatness does not need the Pushpak Viman to be literal. It needs the Isha Upanishad to be understood.

## "We Followed What Others Made"

And there was an answer that came quietly, usually from younger people, often in a mixture of Hindi and English. We just follow others. Humein exposure nahi mila — we never got the exposure. We follow the naksha — the blueprint, the stereotype — that other people drew for us. The syllabus someone else wrote. The career ladder someone else built. The definition of success someone else decided. We are good at climbing — but the ladders are never ours.

This answer holds two griefs at once. The first is the grief of exposure denied — and it is real. Invention needs proximity: to laboratories, to capital, to mentors, to the sheer permission of an environment where trying and failing is survivable. For most Indians across most of the last two centuries, that proximity did not exist. The second grief is deeper, and the young people who said naksha had put their finger precisely on it. Following a blueprint drawn by others is not just a career problem. It is Matsarya (comparison) institutionalized — an entire life navigated by someone else's map. And a person navigating someone else's map never has a reason to consult their own.

Yet notice what this answer assumes: that the map must come from outside. Everything this book has argued says otherwise. The receiver is within. The knowledge is within. Eklavya was refused entry to the official school —

denied exposure in the most literal sense — and went into the forest and drew his own naksha before a clay statue. The absence of exposure is a real wall. But India's own deepest teaching is that the wall surrounds the institution, not the knowledge.

## "Now Everyone Knows Everything"

This answer came from the youngest people I asked. The ones who had grown up with smartphones. With YouTube. With Wikipedia in their pockets. With Artificial Intelligence (AI) assistants that could write code, summarize articles, and pass professional exams.

Now everyone knows everything. There is no advantage anyone can have. Whatever you might invent, someone else somewhere else is already working on. Whatever you might write, someone has already written. The age of individual brilliance is over. We are all just nodes in an information network now.

The despair in this answer is real. And in one sense, it is correct. The age of information scarcity is over. The age of individual differentiation through what you know is ending.

**But in another sense, this is the most encouraging condition India has been waiting for.**

Because if the age of knowing is over, what comes next? The age of being. The age in which the quality of the mind that processes the information matters more than the information itself. The age in which the stillness behind the activity becomes the differentiator. The age in which the technology of consciousness — five thousand years of Indian work on exactly this — becomes the most relevant technology in the world.

Everyone feels behind. Everyone feels that others are growing faster. Everyone is in a hurry that prevents the stillness in which genuine insight can occur.

And here is the particular irony of this moment: in the age of AI, when everyone has the same information, the only thing that differentiates one person from another is not what they know — it is who they are. The quality of their thinking. The depth of their understanding. The stillness that allows them to hear what the noise of the information age obscures.

**The age of everyone knowing everything is the age that most demands the one thing that cannot be downloaded: a quiet mind.**

## What the Mirror Shows

Every answer I received — lazy, no knowledge, colonization, caste, family pressure, the defensive list, the ancient glory, the borrowed blueprint, everyone knows everything — was a mirror.

**Even Indians have forgotten what India was.**

Not because they are not intelligent. Not because they do not care. But because the colonization of the imagination — the deep, persistent belief that the modern industrial model of progress is the only model — has shaped what questions they ask and what answers they consider possible.

The question "why did Indians not invent anything?" assumes that invention — in the modern industrial, patent-and-market sense — is the measure of a civilization's worth. And every Indian who has accepted this assumption — who has answered the question on its own terms, rather than questioning the terms themselves — has, in that moment, conceded the most important point in the argument.

This book is about questioning the terms. When you ask not just why Indians did not invent, but what invention is for — and whether there might be other ways of addressing the same fundamental human needs — a completely different picture of India emerges. Not a civilization that failed to invent. A civilization that chose a different project.

## The Doctor Without a Degree

There is one more observation about Indian culture that I want to make here — the observation that opened Chapter One, returning here from a different angle.

**Every Indian gives health advice.**

You mention a headache, and before you finish the sentence, someone has a remedy. You describe a stomach problem, and three people immediately know exactly what to do. Nobody has a medical degree. But everybody has an answer.

Where does this confidence come from? It comes from something real. It comes from the residue of a civilization that, until a few generations ago, was a continuous transmission of medical knowledge from mother to daughter, from grandmother to grandchild. The kitchen as clinic. The home as the first hospital. The knowledge of which spice for which condition, which preparation for which season, which remedy for which symptom — all carried in the daily practice of every household.

That knowledge has thinned considerably. The grandmother who could diagnose and treat a dozen common ailments has often not transmitted that knowledge to her granddaughter, who grew up in a city, in a nuclear family, with a pharmacy on the corner. But the confidence remained. The pattern remained. Every Indian still gives health advice — not always with the depth that earlier generations had behind it, but with the same instinct that the answer was supposed to be in the family, not the institution.

**The container survived. The contents have thinned. Restoring the contents is what every chapter of this book has been pointing toward.**

## The Most Honest Answer

There was one more answer I received, more rarely than the others, that struck me with particular force.

It came after a long pause. After the person had given the standard answers — lazy, no knowledge, colonization, caste — and seen that I was not satisfied. They sat for a while. And then they said:

*I don't know. I have never really thought about it. I have only heard the question and tried to answer it. I have never asked it myself.*

**This is the most honest answer. And the most hopeful one.**

Because what this Indian was admitting was that the question had been imposed from outside. Asked by foreigners, by media, by the educated elite who had absorbed the foreign frame. Indians, in the deepest part of themselves, had never asked themselves why they had not invented. Because they had never accepted that the absence of invention was the failure that the question assumed it was.

The most honest Indians did not have an answer to the question because they had never accepted the question as legitimate in the first place. They were sitting under the tree, or in the kitchen, or at the temple, or with their family — being Indian in a way that had nothing to do with whether they were inventing. The question was the wrong question. And the most honest answer was the silence in which the wrong question dissolved.

This is the answer this book has been building toward. Not a defence of India against a hostile question. Not a list of what India did invent. But the quieter, more confident position that the question itself was a category error. That India was always doing something different. And that this something different — once understood — turns out to be exactly what the modern world is starting to search for.

*And now we come to the question this book must ask in its final chapters. If India was not failing to invent — if India was building something deeper, slower, and more durable — then what, exactly, would survive if everything else ended? If tomorrow was the last day of this civilization, what is the one thing India would want to pass forward? In the next chapter, we look directly at that question.*

CHAPTER THIRTEEN

# The Last Day

*What Will You Leave Behind?*

*Ram se bada hai Ram ka naam.*

*The name of Ram is greater than Ram himself.*

— *Tulsidas, Ramcharitmanas (sixteenth-century retelling of the Ramayana)*

## The Question at the End of Everything

Imagine the scientists gathered. Not in any one building, but across the world, connected by whatever communication systems still function. They know that something is ending. They have perhaps days. Perhaps years. But the direction is clear.

What do they preserve? What do they encode for whoever comes next?

They would think first of their data. Their equations. Their theoretical frameworks. How do we store the information? How do we make it survive?

And then, perhaps, one of them would ask a different question.

**What if the information is not what matters most? What if the most important thing to preserve is not data — but the capacity to understand it? Not the equations — but the quality of consciousness required to read them and know what they mean?**

India answered this question thousands of years ago. And the answer it chose changed everything about what it built, what it preserved, and what it left for those who came after.

## The Problem Every Civilization Faces

Every civilization that has ever existed has faced the same fundamental problem: how do you preserve what you know across time? Not for one generation. Not for ten. But across hundreds of generations, through wars and invasions and natural disasters and the simple entropy of time that erodes everything built from stone or written on paper.

The Library of Alexandria — which contained, by some estimates, the accumulated knowledge of the ancient world — burned. The libraries of the Maya were systematically destroyed. The knowledge traditions of countless cultures were extinguished by colonization, by forced conversion, by the deliberate erasure of the conquered by the conqueror.

India was invaded. Repeatedly. Its temples were destroyed. Its institutions were dismantled. Its languages were suppressed. Its knowledge traditions were dismissed and derided.

**And yet — the knowledge survived.**

The Vedas (ancient sacred texts of knowledge) survived. The Upanishads (philosophical wisdom texts) survived. The Ayurvedic (ancient health science) texts survived. The stories survived. The rituals survived. Not primarily through stone inscriptions or preserved manuscripts. It survived because India developed the most sophisticated knowledge-preservation system ever created by any civilization on earth. And that system was not made of stone or paper or silicon.

**It was made of human beings.**

## The Living Library

The Rigveda (oldest of the four Vedas) — one of the oldest surviving texts in any language — was composed approximately 3,500 years ago. For approximately 2,000 of those years, it was not written down at all. It was memorized. Transmitted orally from teacher to student, generation after generation, across two millennia, with a precision that modern textual analysis has found to be extraordinary.

Students learned to recite the text not only forward but backward — word by word, phrase by phrase, in both directions. They learned it in multiple recitation patterns, each one a different cross-check on the integrity of the transmission. The result was a redundant, self-correcting memory system that could detect and correct errors with greater reliability than many written traditions.

**Modern computational linguists who have analyzed the Vedic (ancient knowledge) oral tradition have found that the phonetic structure of Sanskrit (ancient Indian language) Mantras (sacred sounds) contains its own error-correction mechanism. This is data compression and error correction — the foundational principles of modern digital information systems — implemented in the human voice and the human memory, thousands of years before the first computer was ever imagined.**

## Zero, One and the Language of the Universe

Every digital system in the modern world — every computer, every phone, every data center, every AI (artificial intelligence) system — operates on a binary foundation. Everything is ultimately encoded as a sequence of zeros and ones.

The concept of zero — without which this binary system would be impossible, without which modern mathematics and therefore modern science and technology cannot exist — was developed in India.

This is not a minor historical footnote. Without zero, there is no place-value number system. Without place value, there is no algebra. Without algebra, there is no calculus. Without calculus, there is no physics. Without physics, there is no engineering. Without engineering, there is no technology. The entire technological civilization that the world has built — including every invention ever cited as evidence of India's backwardness — rests on a mathematical foundation that India provided.

But why did India develop zero? The concept of Shunya (emptiness/void — the absence that contains all potential) was central to Indian philosophical and spiritual thought long before it became a mathematical symbol. Indian meditators had been investigating the nature of emptiness for thousands of years before mathematicians needed a symbol for it.

*The mathematical zero was the philosophical Shunya (void) given a number. And the Purnam (wholeness) of the shloka (verse) that opens this book — that which contains everything — is its complement. The universe of zeros and ones that runs modern technology rests on the same binary that India encoded into its deepest philosophy long before there was any technology to apply it to.*

## The Night I Learned That Forgetting Is Not Human Nature

I want to tell you how I came to see this — because it did not come from a Sanskrit text. It came from an American magician.

First, an honest confession. For much of my life, my family considered my memory weak — and so did I. Send me to the market for five items, and I would return with three. The names, the lists, the small daily instructions — they slipped through me. Years later I understood what was actually happening: my memory was never broken. My attention was simply elsewhere — a mind crowded with problems it was trying to solve has no hands free to carry a list. What everyone called a weak memory was an occupied one. But at the time, I only knew the verdict: he forgets.

Then, in 2015, I read a small book by a stage performer famous across America for a single astonishing feat: he would meet hundreds of strangers at a gathering, once, and then call out every name without an error. His secret was not a gifted brain. It was a system — one he taught openly. Every number, every name, every abstract thing is converted into a word that sounds like it and can be seen as a picture; and the pictures are then bound together into a story, the more vivid and absurd the better. The mind, he explained, forgets abstractions helplessly — but it cannot forget an image, and it cannot lose an image that a story has chained to another.

I recognised the principle at once — because, without knowing its name, I had already used it. In my nursing studies, the terminology had arrived in English, a language then still heavy on my tongue. Pancreas refused to stay — until I stopped repeating it and built it: a paan shop, its counter split by a deep crease. Paan-crease. It never left. Inside the pancreas sit the Islets of Langerhans — and my mind, reaching for both its languages at once, saw them instantly: hans — swans — floating on tiny islands, being served at a langar. Absurd images, ten seconds each, permanent. And notice what happened there: an Indian mind used a paan shop and a langar and a Hindi swan to conquer European medical Latin. If you carry two languages, you carry twice the sounds to build with. The bilingual mind, so often treated as a burden in the classroom, is a mnemonic superpower.

But the discovery went one level deeper. Around that paan shop, over the months, I built the entire subject. The anatomy of the pancreas sat on its counter; the physiology ran behind it like the shop's daily business; the pathologies were the troubles that could strike it; the radiology images hung on its wall like framed photographs. When an examination asked, I did not recall facts — I walked back into the shop and read them off its shelves. It never faded, because it was not a word I was memorising; it was a place I could return to. Chapter Fourteen of this book describes five temples in which India consecrated the five elements themselves into places a human being could walk into — knowledge stored as location, because location survives what repetition cannot. My paan shop was the same technology, at the smallest possible scale. And one more permission, student to student: do not make your images dignified. The mind grips the vivid, the absurd, even the crude, far harder than the polite. Memory does not reward good taste. It rewards full attention.

And the method scales past single subjects. Take five items a family might send you for — milk, soap, bananas, matchsticks, salt. Do not repeat them; see them: a cow showering in soap-foam, peeling a banana whose inside turns out to be a lit matchstick, which you put out by pouring a mountain of salt on it. One ridiculous scene, ten seconds of full attention — and the list cannot leave you, because it is no longer a list. It is a story your mind refuses to lose. The boy who came home with three items was never missing memory. He was missing a scene.

Something in me changed permanently. Memory, I understood, is not a talent that some receive and others lack. It is a method — and attention is its fuel. I never went further than such play; that is the point. The confidence alone rearranged me. I have learned languages and professions since, the ordinary way, through work — but I carry the system like a sealed instrument, knowing that the day I truly need to remember everything, the method is waiting. Forgetting, I learned that night, is not human nature. Unstructured storage is.

Years later, working on this book, the realisation arrived like a key turning. Image instead of abstraction. Story binding image to image. Absurdity as glue. Rules governing the chain. I had seen this system before. It is the compositional grammar of the entire Indian tradition — a god with an elephant's head, an ocean churned with a mountain for a rod and a serpent for a rope, a monkey carrying a mountain because he could not identify one herb.

And then I learned that the resemblance is not a resemblance at all. The magician's number-system works by mapping each digit to consonant sounds, so that numbers become words and words become pictures. India built exactly this device — and built it first. It is called Katapayadi: the consonants of Sanskrit mapped to the ten digits, so that any number can be spoken as a word, a name, even a line of meaningful verse. The astronomers of Kerala used it for centuries to carry their most precious knowledge — the constants of the heavens — as poetry the tongue could not lose. India's system is attested more than a thousand years before Europe produced its equivalent. The modern world's finest memory performer, optimising for what the human mind can actually hold, had arrived at a truth India had already engineered twice over — once for its astronomers, in Katapayadi, and once for everyone, in its stories. He used it to remember names at parties. India used it to remember itself. And I can prove the method on the very name you just met. See it: you cut open a papaya — and inside, in place of seeds, the ten digits gleam like black pearls. Cut-a-papaya-digits: Katapayadi. You will carry the name of India's number-alphabet for the rest of your life — and you have just watched the method work on you.

So let me say directly what this section exists to say — especially to every student reading this who has been told, or has told themselves, that their memory is weak. It is not. There is no such thing as a weak memory; there is only unstructured storage and occupied attention. Your tradition knew this so thoroughly that it wrapped everything worth keeping in images and stories — and if you grew up on those stories, your memory has already been trained by the finest system ever built; you have simply never been told that it was one. Convert what you must keep into pictures. Bind the pictures with a story. Give the story your full attention. That is the entire secret — the magician's, the astronomers', the Rishis'. It was never a superpower reserved for the gifted. It is a method, sitting open, waiting for the day you decide to pick it up.

## Stories — The Original Data Compression

Return to the imagined scientists at the end of everything. What format would they choose for the knowledge they want to preserve?

They could write it down — but paper burns, hard drives fail, satellites fall out of orbit. They could carve it into stone — but stone weathers, and the language carved into it stops being read. They could encode it into DNA — but who, on the other side of the ending, would know how to decode it?

India chose a different format. India encoded its highest knowledge in stories.

**Not as decoration. As the most sophisticated data compression and transmission technology ever invented for human knowledge.**

A story can be told to a child who cannot read. It survives illiteracy. It survives the destruction of every library. It survives the death of every scholar. As long as one grandmother is alive to tell it to one grandchild, the story continues. And the knowledge encoded inside the story continues with it.

This is why the Ramayana (epic of Rama) and the Mahabharata (great epic) and the Puranas (encyclopedic texts) are not simply religious literature in the way the modern category of religion would categorize them. They are the encyclopedias of a civilization that knew it would face invasions, occupations, and the deliberate erasure of its written records — and built a knowledge system that could survive all of these things.

The Indian sages who composed these stories — or who received them and shaped them into the forms in which they have been transmitted — understood something profound. If you encode knowledge in a textbook, you depend on the institution that owns the textbook. If you encode it in a story that every child wants to hear, you depend on no institution at all. You depend only on the fact that human beings love stories more than they love textbooks. And as long as that is true, the knowledge survives.

*This may have been a deliberate choice made by a civilization that saw the cycles of destruction coming. Or it may have been the natural choice of a civilization that simply understood, from long observation, that stories outlive everything else. Either way, the result is the same. The knowledge survives — embedded in narratives whose surface meaning has kept generation after generation entertained while the deeper meaning waits, patient, for the listener who is ready to hear it.*

## The Ramayana's Deepest Layer

To see how much a story can carry, consider the architecture of the Ramayana at its deepest level.

Rama — the embodiment of Dharma (right action), the soul aligned with its highest nature — is separated from Sita, who is not merely a queen but Shakti (feminine creative energy), the principle that gives form to consciousness. And who separates them? Ravana — not simply an evil king, but Raag in its most sophisticated form: the Ahankar of the supremely gifted. Ravana is a master of the Vedas, a devotee of extraordinary discipline, a being of immense knowledge and power — who places all of it in service of ego rather than truth. He is the precise portrait of the ninth face of Raag that Chapter One described: the ego born from knowledge itself.

When Raag separates consciousness from its creative energy, the story says, the world goes out of balance. And the key to restoring the balance is Hanuman — the son of Vayu (air), the master of the medium through which all signals travel, the one who had forgotten his own power until Jamvant reminded him.

Read this way, the Ramayana is not only a narrative. It is a complete diagnosis, encoded in characters: what Raag at its most brilliant looks like, what it steals, and what it takes to win it back — devotion, the mastery of the inner medium, and the remembering of a power that was never lost.

And the tradition even gave us the image of what the restored balance looks like. In the perfected order the story points toward — the reign the tradition remembers as its ideal — the cow and the lion drink water at the same ghat, side by side. Predator and prey at one shore, thirst answered without fear. It is not a claim about zoology. It is the picture of a world in which Raag has been mastered so completely that even the oldest hunger on earth waits its turn peacefully. Every generation that heard the story carried that picture — the measure against which every kingdom, every household, every heart could be checked.

A child hears an adventure. A seeker, returning to the same story decades later, finds the map. The same text, both times.

## When the Sages Saw the End Coming

There may have been a moment in the deep past — not one moment but many, perhaps, across many cycles — when the most awake human beings of an age looked around and saw what was happening. They saw the civilization filling with Raag (desire/attachment). They saw the powerful using their power against each other. They saw that the open transmission of knowledge — the giving of every key to anyone who asked for it — had become a danger. The wrong hands now held tools they would misuse.

And these sages, the tradition holds, made a decision.

They would no longer give the knowledge openly. They would encode it. Hide it inside stories. Embed it in rituals whose surface meaning would be obvious and whose deeper meaning would be invisible to those not prepared to see it. Build it into the calendar so that even an illiterate person would carry the knowledge in the rhythm of their year. Build it into the fasting practices, the Puja (worship) rituals, the daily routines — so that the body would carry the knowledge even when the mind had forgotten what it was carrying.

**This is one possible reading of why so much of India's deepest knowledge is encoded rather than stated openly. And there is a particular dimension of it that any honest observer of Indian life will notice.**

Even an illiterate Indian woman in a village — someone who has never received formal education, who cannot read or write — typically knows the Hindu calendar in detail. She knows which Tithi (lunar day) it is. She knows which Vrata (fast) is to be observed on which day. She knows the rituals appropriate to each season, each festival, each life event. She follows it with precision. She passes it to her daughter. And yet, in many cases, that same woman has never been taught the financial system that governs her household's money — and may not understand it for her entire life.

What does this tell us? The encoded knowledge — the Indian calendar, the seasonal practices, the ritual structure — has reached every layer of society, including those who never learned to read. The non-encoded knowledge — the modern financial system — has not. The Rishis (sages) who encoded the knowledge of nature, time, and human existence into the stories and the calendar and the rituals built better delivery systems than any modern educational institution has yet built. They reached the entire population. Including those that the institutions never could.

## The Examination That Receives the Key

If the deepest knowledge is encoded in stories — if it sits inside the Ramayana and the Bhagavad Gita (divine song) and the Puranas, waiting to be decoded — then who decodes it?

This is the dimension of Indian knowledge that the modern world has had the most difficulty with. Because the modern world believes that if knowledge is good, it should be available to everyone equally, at the same time, in the same form. The egalitarianism of information.

India did not believe this. India believed that knowledge of certain kinds — particularly the knowledge that gives power — should only be received by those who have already done the inner work to handle it without being destroyed by it. The Mantra (sacred sound) given to a disciple unprepared for it is, in the Indian understanding, like giving a precision instrument to a child. It does not bless. It harms.

**So the knowledge was encoded in such a way that it could only be received by those who had reduced their Raag (desire) sufficiently to receive it. The seeker who came to the story with a calm mind, a quiet ego, and a sincere question would find layers of meaning that the seeker who came with the desire to use the knowledge for personal advantage would never see.**

This is what the Indian tradition calls Lila (divine play). The universe itself is structured as an examination — not a test of intelligence or skill, but a test of internal preparation. The doors open for those who have made themselves capable of walking through them. The doors remain closed for those who have not — not as punishment, but as protection. Of them. Of others. Of the knowledge itself.

The stories of the Ramayana and the Mahabharata work in exactly this way. Read at one level, they are extraordinary narratives — heroes, wars, love, betrayal, devotion. Read at another level, they are systematic teachings about Dharma, about the nature of consciousness, about the architecture of the universe. Read at a still deeper level, they are technical manuals for the specific challenges every conscious human being will face on their own arc of development. The same text, different levels, available simultaneously.

India ran two doors at once. The practices — Yoga, Pranayama, Dhyana, the stories, the calendar — were given freely to everyone, without fee and without gatekeeper. The power-bearing keys — specific Mantras, specific ritual technologies — were guarded, given only to the prepared. This is not a contradiction. It is a design: the medicine was free, and the pharmacy's strongest compounds required a physician. And the same honesty applies to the woman who carries the entire ritual calendar but was never taught the financial system. The calendar reaching her proves the delivery system's genius. The financial system not reaching her marks that system's limit — the encoded channel carried what the tradition put into it, and nothing it did not.

And the tradition knew one more thing about transmission — something so honest it is startling: that learning itself is dangerous. Knowledge passing between two human beings creates the perfect conditions for Raag — the student's envy of the master, the master's fear of being surpassed, the ego rising on both sides of the lesson. This book has already shown where that danger leads: a Dronacharya asking for a thumb. So the tradition did something no modern university does. It began every session of study with a prayer against the danger of studying:

ओम् सह नाववतु । सह नौ भुनक्तु ।

सह वीर्यं करवावहै ।

तेजस्वि नावधीतमस्तु मा विद्विषावहै ।

ओम् शान्तिः शान्तिः शान्तिः ॥॥

— *Shanti Mantra of the Krishna Yajurveda*

*May we be protected together. May we be nourished together. May we work together with great energy. May our study be filled with light — and may we never harbor enmity toward one another. Om, peace, peace, peace.*

Read the last line again. Teacher and student, at the threshold of every lesson, acknowledging out loud that the lesson could turn them into enemies — and asking, together, that it not. A civilization that opened its classrooms with this prayer understood the transmission of knowledge more deeply than any civilization that opens its classrooms with attendance.

## Sleep, Memory, and Brahma Muhurta

Modern neuroscience has confirmed something that every grandmother in India has always known: sleep is when memory consolidates.

During sleep — particularly during the deep, slow-wave stages and the REM (Rapid Eye Movement) dream stage — the brain replays the experiences of the day, processes them, and transfers them from short-term to long-term memory. The Hippocampus (the brain's memory consolidation center) communicates with the cortex (the thinking brain), weaving new experiences into the existing fabric of stored knowledge.

India understood this not as neuroscience but as practice. The importance of sleeping and waking at the right times. The practice of reviewing the day's learning before sleep. The ancient prescription to rise before dawn — the

Brahma Muhurta (the creator's hour, approximately 96 minutes before sunrise) — when the mind is freshest, when the consolidation of sleep is most complete and the clarity of a newly organized memory is most available.

**And the grandmother who tells her grandchildren stories before sleep — the stories of Rama (the prince of Ayodhya) and Hanuman (the devoted servant), of the Pandavas (five royal brothers) and Krishna (the divine guide), of the clever jackal and the foolish lion — is using this technology with extraordinary sophistication. The story enters the mind just before sleep. Sleep consolidates it. The child wakes with the story woven into their memory more deeply than anything they will learn in a classroom.**

## What India Chose to Leave Behind

Return now to the question of the last day. What does a civilization choose to preserve when it understands that preservation is the most important work it can do?

India's answer, revealed across thousands of years of deliberate, sophisticated encoding, was this:

**Not the information. The intelligence required to receive it.**

**Not the equations. The consciousness in which equations become understanding.**

**Not the technology. The wisdom to know what technology is for.**

It preserved this by encoding it in every layer of daily life. In the stories that children heard before sleep. In the Ram Naam (name of Ram) that was chanted at every transition. In the Puja (ritual) that began every significant act. In the calendar that structured the year. In the practices that structured the day. In the philosophy that explained the structure of existence itself.

And it preserved all of this not in stone — though stone was also used — but in the most durable medium ever discovered for carrying human knowledge: the living human being, in relationship with other living human beings, across the unbroken chain of generations.

## The Decoding That Is Still Happening

The stories India preserved are not finished giving what they contain.

The understanding that Ram se bada hai Ram ka naam — that the vibrational frequency of a name can be more fundamental than any physical form — connects to modern physics' understanding that at the deepest level of reality, everything is vibration. As Chapter Seven of this book explored, the Indian Rishis treated voice and Mantra with the care that modern physics now recognizes — that an electromagnetic signal, once broadcast, propagates outward and does not stop.

The understanding that the Bhagavad Gita provides — of Karma Yoga (selfless action), of the Atman (soul) beyond the body, of Dharma aligned with the deepest nature of things — is perhaps the most urgently needed framework for human decision-making in the age of AI, when the consequences of actions extend beyond any individual lifetime.

And the understanding that Ganpati (the remover of obstacles, the principle of clearing) must precede every significant beginning — discussed in Chapter Three of this book — is the most practical possible response to the age of AI, in which everyone has access to the same information and the only advantage is the quality of the mind that processes it.

India encoded its highest science in stories. And those stories are still being decoded. By scientists. By philosophers. By ordinary human beings who encounter them and feel, somewhere beneath the surface of the narrative, that something true is speaking.

*India did not wait for the last day to decide what to preserve. It spent five thousand years building the most robust, most multilayered, most human-centered knowledge preservation system ever created. And then gave it, as freely as the Ram Naam is given, to anyone willing to receive it.*

*In the final chapter before the conclusion, we will turn from what India preserved to what India knew about the future. Not as prophecy. As the logical conclusion of five thousand years of understanding how the universe works. The blueprint for conscious invention — the kind that serves life rather than depleting it — has been available for a very long time. The world is only now becoming ready to read it.*

CHAPTER FOURTEEN

# Panchabhuta

*The Blueprint for Future Invention*

*The Atman alone is my Guru. Yet I have gained wisdom from twenty-four teachers who are all around me.*

*— Lord Dattatreya, Srimad Bhagavatam (encyclopedic wisdom text), Canto Eleven*

## The Man Who Learned from Everything

There is a figure in Indian tradition who represents, more completely than perhaps any other, the philosophy that this book has been building toward.

His name is Dattatreya. He is understood in the tradition as the simultaneous incarnation of Brahma (creator), Vishnu (preserver), and Mahesh/Shiva (transformer/dissolver) — the three fundamental forces of the Trimurti (holy trinity) — in a single human form. He is depicted with three faces and six hands. The four dogs (Svan) that accompany him represent the four Vedas (ancient sacred texts of knowledge). The cow standing behind him represents Kamadhenu (the wish-fulfilling cow that emerged from the Samudra Manthan, discussed in Chapter Four).

One day, a king named Yadu encountered Dattatreya wandering in a forest — laughing, weeping, apparently unconcerned with anything the world considers important. The king, struck by the sage's evident happiness in the absence of every conventional source of happiness, asked him:

*Who is your Guru (teacher)? Who taught you to be this free?*

Dattatreya's answer is preserved in the eleventh canto of the Srimad Bhagavatam, and it is one of the most extraordinary statements about the nature of knowledge and learning that any tradition has ever produced:

*The Atman (soul/self) alone is my Guru. Yet I have gained wisdom from twenty-four teachers — from the earth, the water, the air, the fire, the sky, and from creatures and human beings in whom I saw the teaching I needed to receive.*

Twenty-four teachers. Not one human Guru with credentials and an institutional affiliation. Twenty-four teachers from the living world around him — most of them not human at all.

**This is, by any measure, the most radical educational philosophy ever articulated. The entire world is a teacher. Every element, every creature, every situation carries a teaching for the consciousness that is awake enough to receive it.**

## The Twenty-Four Teachers

The first five teachers are the Panchabhuta (five elements) themselves:

Prithvi (Earth element) — taught patience and endurance. The earth bears everything done upon it — every wound, every weight, every abuse — and continues to nourish. It gives without keeping account. It sustains without condition.

Jal (Water element) — taught purity and flow. Water purifies whatever it touches without becoming impure itself. It moves constantly. If it stops, it stagnates. Keep moving — this is the lesson of water.

Vayu (Air element) — taught freedom from attachment. Air touches flowers and thorns alike, carries every fragrance and every stench, yet attaches to nothing. It moves through the world without taking the world with it.

Agni (Fire element) — taught transformation. Fire consumes whatever is offered to it and produces only warmth and light. It does not discriminate between the offering of a sage and the offering of a fool — it transforms both into the same heat.

Akasha (Space element) — taught infinite capacity. Space holds everything and is touched by nothing. It contains every form and remains itself empty, free, available.

**These five teachers — Earth, Water, Fire, Air, and Space — are not just elements of the material world. They are the first and most fundamental teachers. The primary school of the universe, available to every human being in every moment, requiring no institution, no degree, no fee.**

The remaining nineteen teachers come from the world of living beings and human situations. A selection of the most important ones:

The Sun — taught that the same light, reflected by different surfaces of water, appears as many suns. The Atman (soul) is one. Its appearance as separate individuals is the reflection. The reality remains one.

The Pigeon — taught the danger of excessive attachment to family. A pigeon, having seen its young captured in a hunter's net, threw itself into the same net rather than live without them. Attachment, even to the deepest love, can become a trap.

The Python — taught contentment with whatever arrives, taking only what comes without struggle. The python does not chase its food. It waits. What comes, it accepts. What does not come, it does not pursue.

The Sea — taught steadiness amid disturbance. Rivers flow into the sea, bringing all manner of debris and tumult. The sea receives all of it without changing its essential nature. The level of the sea remains. The waves come and go on its surface.

The Moth — taught the danger of being attracted by what destroys. The moth, drawn to flame, perishes in it. Raag (desire/attachment), in any of its forms, leads into traps.

The Deer — taught the danger of being distracted by beauty. The deer, enchanted by music, is caught. Aesthetic attachment, like all attachment, has its shadow.

The Fish — taught the danger of appetite unchecked. The fish swallows the hook hidden in the bait. Every attachment has its hook.

Pingala — taught that attachment to the expectation of receiving something destroys peace. When she finally gave up hoping and rested in acceptance, she found the peace that had eluded her while hoping.

The Child — taught freedom from past and future. The child lives completely in the present, unconcerned with yesterday's insults or tomorrow's needs. Pure Purnam (completeness).

The Young Girl — taught solitude as the condition for undistracted work. Pounding rice alone to avoid the noise of guests who broke her rhythm.

The Arrowsmith — taught complete absorption in work. So focused on straightening his arrow that a royal procession passed without his awareness. This is the state in which the greatest work happens.

The Serpent — taught solitude and the value of moving without fixed home. The serpent lives alone, moves silently, takes shelter where it can. No entanglement. No permanent claim.

The Spider — taught the cyclical nature of creation. The spider spins its web from within itself and then absorbs it back. Creation from within, dissolution back into the self. The universe does this continuously.

The Wasp — taught the power of focused attention to transform. The caterpillar, thinking only of the wasp, becomes the wasp. Whatever you fix your attention on completely — you move toward becoming. As a person thinks, so they become.

## The Teaching Within the Teaching

Look at this list of twenty-four teachers and ask: what kind of education produces a human being who can learn from a moth? Who can receive the teaching of a python? Who can look at a pigeon's tragedy and hear a lesson about their own attachment?

It requires a consciousness that is not preoccupied with its own Raag. A mind that is still enough to receive. A person who has done enough of the examination that the noise of their own desires and fears does not drown out what the world is trying to teach.

**Notice: the first five teachers are the Panchabhuta themselves. Earth. Water. Air. Fire. Sky. The five elements are not merely the building blocks of the physical world. They are the first and most fundamental teachers. The primary school of the universe.**

## The Five Elements and the Universe They Describe

The Panchabhuta — Prithvi (Earth), Apas or Jala (Water), Agni (Fire), Vayu (Air), and Akasha (Space or Ether) — is described across the oldest texts of the Indian tradition. The Taittiriya Upanishad (philosophical wisdom text) traces the order of creation through these elements: from Space came Air, from Air came Fire, from Fire came Water, from Water came Earth. Each element arises from the one before it, each more dense, each more tangible, each more visible to the ordinary senses.

But this is not merely a cosmological origin story. It is a map of how reality is structured. Because what the ancient Rishis (sages) understood — and what modern physics is only now beginning to confirm — is that the material universe, with all its apparent complexity, rests on a hierarchy of states, from the subtlest vibration of space to the densest arrangement of solid matter. The five elements describe that hierarchy with extraordinary precision.

Consider what modern science knows about the human body. Approximately seventy percent of it is water. Another significant portion is earth — solid matter, minerals, structure. A measurable portion is air. A portion is fire — the metabolic energy, the temperature regulation, the chemical combustion that keeps every cell alive. And the remaining is space — the spaces within cells, within bones, within the architecture of the body that allow it to move, breathe, and exist.

**These are the same proportions and categories that Indian philosophy has used for thousands of years. Not as metaphor. As precise description.**

## The Pancha Bhoota Sthalas

India did something with the Panchabhuta that no other civilization has done. It built five temples, distributed across the southern peninsula, each consecrated to one of the five elements — each one designed so that a human being could directly experience that elemental principle in a specific place. Together they are called the Pancha Bhoota Sthalas.

Ekambareswarar Temple in Kanchipuram is consecrated to Earth — Prithvi. Jambukeswarar Temple at Thiruvanaikaval is consecrated to Water — Apas, its inner sanctum built around a living underground stream. Arunachaleswarar Temple at Tiruvannamalai is consecrated to Fire — Agni, and the hill of Arunachala itself is honoured as a living Agni Lingam. Sri Kalahasti Temple in Andhra Pradesh is consecrated to Air — Vayu, where a lamp inside the sanctum flickers though the air is still. And the Chidambaram Nataraja Temple — one of the most architecturally extraordinary structures in the world — is consecrated to Space itself, Akasha, the subtlest of the elements, the one from which all the others arise. Its innermost shrine famously enshrines the Chidambara Rahasyam — the secret of Chidambaram: behind the curtain, there is no image at all. Empty space, garlanded and worshipped. Akasha, presented as itself.

These were not merely places of worship. They were research stations of a different kind — places where the relationship between the human being and the elemental structure of the universe could be felt, studied, and integrated, not through instruments but through direct experience, generation after generation, in the same consecrated conditions. India built its laboratories in stone. And consecrated them for eternity.

## Inventing With the Elements

And the history of invention is, in large measure, the history of a civilization's relationship to these elements — whether it worked with them or against them.

Every invention that works with the elements tends to be sustainable. Its excretion feeds the next cycle rather than poisoning it. Every invention that works against the elements tends to be self-destructive. It produces the Halahala (poison) before the Amrita (nectar) — exactly as Chapter Four warned. And if nobody is prepared to hold the poison, the system fails.

## The blueprint for future invention is not new. It does not need to be invented. It is already encoded in the five elements:

Prithvi teaches: be patient. Build for the long cycle. Give more than you take. Sustain what you begin.

Jal (Water) teaches: stay pure. Keep moving. Distribute what you receive. Flow toward the lowest ego point — toward Seva (service) rather than accumulation.

Vayu teaches: do not attach. Touch everything with freshness. Carry what you carry without becoming it.

Agni teaches: transform rather than destroy. Consume what is given and produce light. Let the excretion of your process be warmth, not toxin.

Akasha (Space) teaches: have infinite capacity without infinite claim. Make room for everything without being confined by anything. Hold the space in which creation can happen.

## Panchabhuta in the Kitchen

Chapter Ten of this book established that the Indian kitchen was always the clinic — the place where preventive medicine was practiced daily, by women who carried in their cooking a body of knowledge refined across hundreds of generations. The Panchabhuta provides the deeper framework for why those practices worked.

Every Indian kitchen was once a complete pharmacy, organized — without anyone saying so — along the five-element framework. Haldi (turmeric — the golden spice) and the root vegetables and grains carry the Prithvi principle into the body — grounding, sustaining, providing structure. The Dal (lentil preparation) cooked with water, the buttermilk, the seasonal drinks carry the Jal principle — flowing, purifying, distributing. The spices that warm the body — Adrak (ginger), Lavang (clove), Dalchini (cinnamon), Kali Mirch (black pepper) — carry the Agni principle, transforming what they enter. The breath practices done before eating, the rhythmic chewing, the conscious awareness of food — carry the Vayu principle. And the periodic Upavasa (fasting) — the days of rest from food entirely — carries the Akasha principle, creating the emptiness in which the body heals.

**The Indian woman who made these preparations was not following a recipe. She was practicing Panchabhuta-aligned preventive medicine. She was doing what Dattatreya did — learning from the earth's abundance, the water's purifying flow, fire's transforming heat — and applying that learning to the most immediate and most important work: keeping her family healthy.**

The future of global health — the sustainable, accessible, non-extractive future that does not bankrupt families or deplete the earth — looks less like a pharmaceutical industry and more like an Indian kitchen. Less like a hospital and more like a grandmother who knows which herb to use in which season.

## Shri Anna — The Ancient Superfood

Millets — which India calls Shri Anna (the auspicious grain) — are among the oldest cultivated crops on earth. They were the foundation of Indian agriculture for thousands of years before rice and wheat became dominant.

And they are, by almost every nutritional measure, exceptional. They are rich in protein, fiber, iron, calcium, and B vitamins. They have a low Glycemic Index (the rate at which food raises blood sugar) — making them ideal for managing blood sugar. They are drought-resistant, requiring a fraction of the water that rice cultivation demands. They grow in poor soils without synthetic fertilizers. They are naturally pest-resistant.

**Modern nutritional science is now calling millets a superfood. Ancient India called them Shri Anna (auspicious grain) and ate them every day.**

This is the Panchabhuta principle in agriculture. A crop that works with Prithvi rather than depleting it. That uses Jal wisely rather than exhausting it. That requires Agni in minimal and efficient processing forms. That gives Vayu nothing to fear — requiring no pesticide clouds. And that contributes to Akasha (sky/atmosphere) health by sequestering carbon rather than releasing it.

The future of agriculture, which modern agronomy is struggling to invent, was already growing in India's fields for six thousand years.

## What Conscious Invention Looks Like

### India is not saying: stop inventing.

India is saying: invent differently. Invent as Dattatreya learned — by listening to the world around you rather than imposing your will upon it. Invent as Prithvi gives — generously, patiently, sustainably. Invent as Jal flows — toward the lowest ego point, toward genuine Seva. Invent as Agni transforms — consuming what needs to be consumed and producing warmth and light, not toxin. Invent as Akasha (sky) holds — with infinite capacity and zero confinement.

Conscious invention asks, before the first prototype is built: What problem does this actually solve? Whose Raag (desire) is it serving? Will its excretion nourish the next cycle or poison it? Is there a way to solve this problem that works with the Panchabhuta rather than against them? Have we learned from the moth and the deer and the fish — from the warnings about what happens when Raag drives without wisdom?

The healer who serves their patient not by waiting for them to be sick, but by working to keep them well through the kitchen, through seasonal practice, through the Panchabhuta principles of Ayurveda (ancient Indian health science) — this is conscious invention. The farmer who returns organic matter to the soil, who maintains the biodiversity of his field, who plants Shri

Anna (millets) alongside his staple crops — this is conscious invention. The engineer who designs a product whose entire lifecycle, from creation to disposal, works with natural systems rather than against them — this is conscious invention. The architect who builds a home whose walls cool themselves through the geometry of cross-ventilation, whose light comes from the sun rather than from electricity, whose materials return to the earth when the building is finally retired — this is conscious invention.

These are examples of the Panchabhuta principle applied to the specific problems of the present moment. And they point toward what all conscious invention — in healthcare, in agriculture, in education, in energy, in communication — can look like when guided by the wisdom that Dattatreya received from the world around him.

## The Future India Already Lived

The world's most urgent problems — climate change, biodiversity collapse, mental health crisis, the meaning vacuum created by AI (artificial intelligence), the inequality generated by Dravya (material wealth)-centered economies, the loneliness epidemic of hyper-connected but genuinely disconnected societies — all have the same root.

**Raag.**

And the solutions — the genuine, sustainable, human-centered solutions — all point toward the same answer. What Dattatreya found in the forest. What the woman preserves in her kitchen. What the farmer knew in his relationship with the soil. What the Rishi (sage) encoded in the story. What the meditator discovers in stillness. What the newborn embodies before the world teaches it to want.

The answer is Purnam (wholeness). The completeness that was the original state. The completeness that remains the natural state once Raag is sufficiently reduced. The completeness toward which every chapter of this book has been quietly pointing.

*India did not fail to invent. India invented the human being. And the human being who has become Purnam (whole) is the one technology the modern world has not yet built — and the one technology, in the age of AI, that matters more than all the others combined.*

*In the conclusion that follows, we return to the question this book began with. We return to the shloka that has been waiting at the very front of the book — the verse from the Isha Upanishad that opened with the word Purnam. We return to the newborn in the arms of the mother in the preface. We return, in the way that India has always taught us to return, to the place we started — knowing it, this time, for the first time.*

CONCLUSION

# The Invention India Was Always Making

*And What It Changes About Everything*

ॐ पूर्णमदः पूर्णमिदं

पूर्णात् पूर्णमुदच्यते।

पूर्णस्य पूर्णमादाय

पूर्णमेवावशिष्यते॥

ॐ शान्ति शान्ति शान्ति

*That invisible universe is whole and complete.*

*This visible world is also whole and complete.*

*From wholeness, wholeness emerges.*

*And even when wholeness is taken from wholeness,*

*only wholeness remains.*

*— Isha Upanishad, opening invocation*

## The Shloka You Now Understand

You have seen this Shloka (verse) twice now.

Once at the very beginning of this book — when it stood alone on a page in silence, untranslated and unexplained. And once here, at the end, after fourteen chapters of journeying through the question that this book was built around.

The first time, it was a mystery.

**This time, it is a recognition.**

That invisible universe from which everything emerges — Purnam (whole/complete). Complete. Whole. Lacking nothing.

This visible world in which we live and struggle and love and invent — also Purnam (complete). Also complete. Also lacking nothing, when seen clearly.

And you — the person reading these words, in whatever moment and whatever circumstance — also Purnam. Born complete. Carrying the knowledge of the universe within you. Simply forgetting it in the noise of Raag (desire/attachment).

*Even when wholeness is taken from wholeness, only wholeness remains.*

तात स्वर्ग अपवर्ग सुख, धरिअ तुला एक अंग।

तूल न ताहि सकल मिलि, जो सुख लव सतसंग।।

*Place all the pleasures of heaven and even liberation itself on one side of the scale. They do not, all combined, outweigh a single moment of Satsang — the company of truth.*

*— Tulsidas, Ramcharitmanas, Sundarkand*

India did not invent because India did not believe anything essential was missing. And when nothing essential is missing, the frantic urgency of Raag (desire)-driven invention simply does not take hold.

## What India Was Inventing

The question that began this book was: why have Indians not invented anything?

We have now traveled through fourteen chapters of answer. And the answer has gradually revealed itself to be a different question entirely.

*What was India inventing?*

**India was inventing the complete human being.**

Not the productive human being. Not the consuming human being. Not the achieving human being. The complete one. The one who has done enough of the examination to hear the signal of their own completeness.

Every chapter of this book has been a different angle on the same project.

Chapter One showed us the disease — Raag — and the human being's ability to release it through the discipline that the first verse of the Ashtanga Hridayam (Heart of the Eight Branches of Medicine) prescribed.

Chapter Two showed us how this discipline was woven into the simplest act of getting dressed. White clothing was not poverty. White was the daily reduction of Raag into the texture of ordinary life.

Chapter Three showed us that prosperity itself is whole — Mahalakshmi (complete prosperity) before Dravya Lakshmi (material wealth) — and that the physics of wealth, like all physics, flows toward the humble.

Chapter Four showed us the warning. Every churning of any depth releases the Halahala (poison) before it releases the Amrita (nectar). The treasures were always already in the ocean. The churning revealed them. It did not create them.

Chapter Five showed us the model. The cow as the complete system. The civilization that contemplated the cow daily was training itself, without saying so, to perceive completeness as the natural condition of any well-functioning whole.

Chapter Six showed us the internal technology. The electricity within you. The Nadi (energy channel) network. The Pranayama (breath regulation) that operates the system. India did not need to invent the external grid because India was running on the original grid.

Chapter Seven showed us India's invisible inventions. Mantra (sound). Yantra (geometry). Tantra (network). Karmkand (encoded ritual). And the recognition, encoded in the Vedas (sacred texts of knowledge) themselves, that the knowledge was heard, not composed.

Chapter Eight showed us the receiver. The human being as the instrument that receives the universal field. Aham Brahmasmi (I am the universal). Brahma Smritva (Brahma remembered). The lemon that proved consciousness produces physical change. The Ram-Saraswati story that revealed even Manthara was acting from the cosmic script. The Taraaju (balancing scale) of Ram and Krishna — two avatars, one Dharma, opposite methods.

Chapter Nine showed us the living proof. Every newborn arriving complete. The Sutika (postpartum) practices that protected the moment when the receiver was at its clearest. Eklavya — who proved that the knowledge was within, accessible to anyone with the patience and devotion to listen for it.

Chapter Ten showed us the body's place in the larger life. The microbiome within us. The earth as a womb. The kitchen as the original clinic. The brain as the original pharmacy. The Navratri (nine sacred nights) practices that timed themselves to the body's seasonal shifts. The Ardhanarishvara (half-male, half-female) reminder that completeness contains both.

Chapter Eleven showed us the longer arc. The Atman (soul) that travels through 8.4 million forms, accumulating the only wealth that survives the change of clothes — the Samskaras (impressions) of action and intention. The cycle that has no end, in which the future generation may be the inventor.

Chapter Twelve showed us the mirror. The answers Indians give when asked the question that began this book — and the colonization of imagination that produced those answers. The honest silence at the bottom of every answer, where the question itself dissolves.

Chapter Thirteen showed us the last day's question. What does a civilization preserve when it understands that preservation is the most important work it can do? India answered five thousand years in advance — not the information, but the intelligence required to receive it. Encoded in stories. Carried in the human voice. Stored in the unbroken chain of generations.

Chapter Fourteen showed us the blueprint. Dattatreya and his twenty-four teachers. The Panchabhuta (five elements) as the most fundamental classroom. The kind of invention that works with the elements rather than against them. The future that India already lived.

*Each chapter was a different angle on the same single project — the invention of the complete human being.*

## Why Now

There is a reason this book has arrived at this particular moment in human history.

Until recently, the project India was working on could be — and often was — dismissed by the rest of the world. The civilizations that built the steam engine and the printing press and the antibiotic and the computer could look at India and say: you may have your philosophy, but we have the things that actually change the world.

That argument is collapsing.

The age of artificial intelligence (AI) — now beginning in earnest — is the first age in human history in which almost everything humans have been uniquely useful for is being done, faster and cheaper, by machines. Writing. Diagnosis. Design. Analysis. Translation. Pattern recognition. Even creative work in many forms. The list of things only humans can do is shrinking, and the rate at which it is shrinking is accelerating.

This produces, across the world, a deep and growing question. If the machines can do almost everything we used to be valued for — what are we for?

**This is the question India has been answering for five thousand years.**

Not because India anticipated AI. But because India never accepted that what humans were for was external production in the first place. India always understood the human as a receiver — a being whose deepest function is the development of consciousness itself, the reduction of Raag, the remembering of Purnam (completeness).

In the age of AI, the things only humans can do — the things that no machine, however sophisticated, can ever do — are exactly the things India spent its entire civilizational arc cultivating. The capacity to be still. The ability to release Raag. The depth of presence that allows a small action to carry the weight of a complete consciousness. The Seva (selfless service) that does not even register itself as service. The Karma Yoga (yoga of selfless action) that releases the impressions of past lifetimes. The Bhakti Yoga (yoga of devotion) that opens the heart beyond what any algorithm can simulate.

These were not luxuries. They were not the activities a civilization pursued because it had finished its real work. They were the real work. And the rest of the world — having spent five hundred years building the external — is now arriving at a moment in which the internal is all that matters.

## The Return Already Happening

The return is already underway.

Yoga (mind-body practice) is being practiced by tens of millions of people in countries that, a generation ago, would not have known what it was. Meditation studios are common in cities that have no historical connection to Indian thought. Ayurvedic (ancient health science) principles are appearing in the recommendations of mainstream physicians. Plant-based eating is being rediscovered as the diet that humans evolved on. Even concepts that until recently sounded mystical — neuroplasticity, the gut-brain axis, the role of breath in regulating the nervous system, the long-term consequences of inflammation — are being researched with the same scientific seriousness that ancient India brought to them by other means.

This is not Indians teaching the world. This is the world arriving, on its own terms and through its own methods, at conclusions India has been holding for a very long time.

And here is the warning that the Indian tradition has always carried: the return itself can be distorted by Raag. Yoga can become a market. Meditation can become productivity hacking. Ayurveda can become a brand. The deep practices designed to release attachment can be marketed in ways that reinforce attachment. This is the Halahala of the current churning. It is real. It must be acknowledged.

**But the signal underneath the noise is unmistakable. Humanity is, slowly and unevenly, returning to what India has been keeping safe.**

And India itself — having spent its colonial and post-colonial period in some confusion about its own value — is beginning to remember what it has. The young Indian who looked at me across a Chai (tea) shop and said "we had no knowledge" is still common. But there is also, increasingly, the Indian who looks at the modern world and sees clearly what is missing, what was always missing, what India was always offering.

And I have watched this return happen in a way statistics cannot capture. In 2015, in the same season the question of this book arrived, ideas began coming to me that I had read nowhere — that the kitchen was the clinic; that every question I asked already carried its answer within me, waiting like the treasures in the ocean before the churning. I wrote them down as my own. Then, years later, I heard a college principal — a man I had never met, who had read nothing of mine — speak the same ideas, in almost the same words. My first feeling was the small sting of ownership. My second feeling was recognition, and it has never left me: of course he said it. The knowledge was never mine. It is not separate from him, or from me, or from anyone raised inside this civilization's stories, festivals, and kitchens. We were not two authors having the same idea. We were two receivers, tuned by the same tradition, picking up the same signal.

This is what the return looks like from inside India: people who have never met beginning, independently, to say the same things. And it points toward something worth imagining slowly. Chapter Two described what happened in a room where everyone wore the same white — the eyes were given nothing to rank, and the arguing lost its fuel. Now imagine a civilization where everyone has remembered — where the shared knowledge is awake in every mind, and everyone speaks, in that deeper sense, the same language. Not the same opinions; the same grammar of understanding. That is not uniformity. That is Satsang at the scale of a nation — the company of truth, held in common. The tradition's word for gathering was never accidental. The knowledge is already distributed; it always was. What remains is only the sitting together that switches it on.

## India Did Not Invent. India Remembered.

This book began with a question that had a hidden assumption.

The question was: why have Indians not invented anything?

The hidden assumption was that invention — in the modern industrial, patent-and-market sense — is the measure of a civilization's worth. Accept the assumption, and India looks like a civilization that fell behind. Question the assumption, and India looks like a civilization that was playing a different game.

And once you understand the game India was playing — once you see that India was inventing the complete human being while the rest of the world was inventing things for the incomplete one — the entire question reorganizes itself.

India did not fail to invent. India also invented many things along the way — zero, the decimal system, plastic surgery, foundational mathematics, more — and gave them freely to the world. But these were not the project. They were by-products. The actual project was the human being.

And the human being India was building was not built through invention in the external sense. The human being was remembered. Brought forward, generation after generation, through stories and rituals and seasonal practices and the daily reduction of Raag, into the form that was always its natural condition.

**India did not invent. India remembered. And remembering, in the deepest sense, is what every human being is here to do.**

## The Reader Who Has Arrived Here

You have read this book. Across many pages and many ideas, you have walked through what one Indian observer has tried to say about the civilization he has watched all his life.

Some of it you agreed with. Some of it you wanted to push back on. Some of it you carried with you for a few hours and then forgot. Some of it, perhaps, will surface again in a quiet moment months from now and mean something then that it could not mean at the moment of first reading.

All of this is exactly as it should be. This was not a book that asked you to agree. It was a book that asked you to consider.

And if anything in these pages has done its quiet work — if even one small place in your understanding of India, or of yourself, or of what a human life is for, has shifted by the smallest measurable amount — then everything the book was trying to do has succeeded.

Because India's project was never about converting anyone. It was about being available. Like the leaf that does not insist the sun see it. Like the lotus that simply opens, regardless of whether anyone is watching. Like the grandmother who continues to make the seasonal preparation even if her grandchildren no longer ask for the recipe.

*The knowledge was always going to be here. Whether you found it or not. Whether the modern world rediscovered it or not. Whether anyone, anywhere, was paying attention or not. The knowledge sits in the field of consciousness, broadcasting continuously, waiting only for the receiver that is quiet enough to hear it.*

You are that receiver. You always were. The book ends here. The remembering does not.

---

There is one word that holds everything this book has been trying to say.

It is the word that began the shloka at the front of the book — when you did not yet know what it meant.

It is the word that has appeared in nearly every chapter — sometimes as a concept, sometimes as a thread, sometimes as a quiet undercurrent beneath the more visible argument.

It is the word that the newborn carries into the world without effort. It is the word that the Yogi (practitioner of union) labors a lifetime to return to. It is the word that the meditator glimpses at the deepest point of stillness. It is the word that the dying person, in the moment before the second umbilical is finally cut, often reports as the texture of the awareness that meets them.

It is the word for what India was always already inventing — never as an external technology, always as the natural condition of the consciousness that has remembered itself.

It is the word with which this book began. And it is the word with which this book ends.

*Purnam.*

# AUTHOR'S NOTE

## On Structure, Choices,

## and What Comes Next

### A Few Words From the Author

Some readers like to know how a book was put together. This short note is for them. The rest of you have my full permission to skip it — the book is complete without it, and what follows is mostly a few quiet explanations of decisions that may otherwise seem strange.

### Why Fourteen Chapters

This book has fourteen chapters. Not ten. Not twelve. Not fifteen. The number is deliberate.

Indian symbolic numerology gives several natural options for the structure of a book on this subject. Twelve, for the twelve Rashis (zodiac signs) of the Indian astronomical year. Fifteen, for the fifteen Tithis (lunar days) of each Paksha (half-month). Eighteen, for the eighteen chapters of the Bhagavad Gita (the divine song).

I chose fourteen for the fourteen years of Vanvas (forest exile) that Lord Ram (the prince of Ayodhya) spent away from his kingdom.

The story of Ram's Vanvas is, among other things, a story about leaving the known and returning to it as someone different. Ram entered the forest as a prince who had not yet been tested. He returned as something that the entire civilization has carried as a model of completeness for thousands of years. The Vanvas was not a punishment. It was an examination. And in passing through it — in encountering Bharadwaj and Atri and Agastya and the long line of sages whose hermitages he visited, in losing what he most loved, in choosing Dharma (right action) under unbearable pressure — he became fully what he had always been.

Every reader of this book is, in some sense, in their own Vanvas. Walking through fourteen chapters of inquiry into a question that does not yield to easy answers. Encountering things that may be uncomfortable. Losing certainties that may have felt important. And, if the book has done its quiet work, arriving at the end of the fourteenth chapter — and the Conclusion that follows — having become, by some small measurable amount, more themselves than they were when they began.

That is what the number fourteen is honouring. Not Ram's literal forest. The structure of every genuine journey through difficult questions.

Why "The Cycle" and Not "The Future"

The third Part of this book is called "The Cycle." An earlier version of the manuscript called it "The Future." The change is small but it carries something important.

Western philosophical traditions have largely understood time as a line. A beginning, a middle, an end. A movement from primitive to advanced. From past to future. The very word "progress" — which the modern world uses without much thought — implies a forward motion along this line.

Indian philosophy understood time differently. Time, in the Indian view, is a cycle. The seasons return. The Yugas (great ages — Satya, Treta, Dwapara, Kali) move in sequence, and when the last one ends, the first one begins again. The cosmos itself, in some Vedic (sacred knowledge) accounts, breathes in and out across timescales so vast that any human lifetime is barely a single breath of a single inhale.

To call the final Part of this book "The Future" would have been to import the linear assumption that the book itself has been quietly questioning. The future implies somewhere new. Somewhere we have not been before. Somewhere we are headed.

But what India was offering was never a future. It was always a return. To the completeness that was the original condition. To the wisdom that was always already here. To the cycle that the modern world had stepped out of and is now, slowly, finding its way back into.

"The Cycle" honours this. It says: what is coming is not new. It is the same wheel, turning, arriving back at the part of itself we had forgotten.

A Few Other Decisions

There are no named living people in this book — only Tulsidas (the sixteenth-century poet-saint whose Ramcharitmanas is one of the great works of Indian devotional literature) is referenced directly, and he died in 1623. This was deliberate. The argument of the book is civilizational. Naming specific modern figures — however illustrative — would have pulled the focus toward individuals when the argument needed it to stay on the larger pattern.

A related choice: this book draws chiefly from the Vedic and Hindu stream of Indian thought, with occasional lines from the Buddhist and Jain traditions where they compress the same insight. India's other great paths — the Bhakti and Sufi confluences, the folk and tribal knowledge systems — carry their own vast answers to this book's question. Their near-absence here is a limit of one author's reach, not a ranking of their worth.

Where modern scientific research is referenced, I have tried to do so with honesty. Some of the research I cite is well-established. Some is preliminary. Some is contested. I have tried not to overstate any of it. Where I have used phrases like "a growing body of research suggests" or "some studies have observed," I mean exactly that — not proof, but the kind of evidence that warrants serious attention. The reader who wishes to investigate further will find some starting points in the Notes and Further Reading that follows.

The Sanskrit and Hindi terms have been translated into English in brackets at first use within each chapter. This is occasionally awkward, but I judged it more useful to the global reader to carry the original word and its meaning together than to lose the precision of the original or to lean too heavily on transliterations alone.

The book does not include footnotes within the main text. The reading experience seemed more important than the academic apparatus. Readers who want sources should turn to the Notes section at the end.

My location is given as "Rajasthan, India" rather than any specific city. This is partly for privacy and partly because the argument of the book is rooted in a place that is older than any specific Indian city — the broader civilization that produced everything this book has tried to describe.

What Comes Next

This book lays out a civilizational argument across fourteen chapters. Some of the arguments deserve, in time, their own full books. Several chapters could, in principle, be expanded into book-length treatments. I am beginning with one of them.

Chapter Six of this book — "The Electricity Within You" — introduced a question that this book had room to raise but not room to fully answer. The question, in its simplest form, is this:

If a few hundred watts of external electricity, channelled through copper wire and silicon, can power a microwave that cooks food in ninety seconds, can run a satellite for decades on solar cells, can carry the voice of one human being to another human being on the opposite side of the planet — what is the electricity within a human being capable of?

The bioelectric system of the human body is, by some measures, one of the most complex electrical systems known to exist. It runs continuously for eight decades on food and breath. It generates the consciousness that is reading these words. It coordinates trillions of cells without a central processor. And humanity has spent five hundred years and trillions of resources developing external electrical systems while leaving this internal one almost entirely unexplored by the modern scientific method.

Indian civilization did not leave it unexplored. For five thousand years, the Yogis (practitioners of union) and the Rishis (sages) and the unbroken chain of teachers and students were investigating exactly what the electricity within a human being can do. Their findings — preserved in the Yoga Sutras (foundational text of yoga), in the Tantric (network-based) traditions, in the documented practices of advanced meditative states, in the reports of capacities that ordinary biology does not predict — are substantial.

Some of what the tradition claims will read, to a modern reader, as fantasy. Some is well-documented in the medical literature. Some sits in a contested middle ground that mainstream science has not known what to do with. The next book will treat all of it seriously — neither dismissing it as superstition nor accepting it uncritically as proven. It will follow the same approach this book has followed: honest with what is known, careful with what is contested, open to what may yet be discovered.

The next book is in preparation. Its working title is the same as Chapter Six of this one:

Electricity Within You

It will explore, in much greater depth, the consume-excrete law that runs through all of nature, the bioelectric system of the human body, and the question of what humanity could give back to the natural cycle in the way that every other living thing already does. And it will take seriously the largest implication of all — that humanity may have been investing in the wrong electricity for five hundred years, and that the internal system, properly understood, may open doors that the external one cannot reach.

If this book has been worth your time, the next one may be worth waiting for. Updates on the project will be available at whyindians.com — the small publishing home of these books — and the related domain that will carry the second book when it is ready.

A Final Acknowledgement

This book carries my name as the author. The honest truth is that what is in it does not belong to me. It belongs to a civilization that has been doing this work, generation after generation, for thousands of years before I was born — and that will continue doing it for thousands of years after I am gone.

Every grandmother who taught a granddaughter which spice to use in the change of seasons. Every father who paused before a meal to acknowledge the food. Every priest who continued to perform the Karmkand (ritual action) when the wider culture had begun to laugh at it. Every Yogi (practitioner) who held a practice that no one else was holding. Every villager who stood under a tree and noticed something most modern people no longer notice. All of them, together, are the actual author of this book.

I am only the one who happened to write it down at this particular moment, when a number of conversations had stayed with me long enough that they finally insisted on becoming a book.

If anything in these pages helped — credit it to the civilization that produced it. If anything seemed wrong — that is mine.

— Raj

Rajasthan, India

# For the Reader

## Who Wants to Go Further

This section is not a comprehensive bibliography. It is a set of starting points for the reader who wants to investigate further the arguments and claims that this book has made. It is organized by chapter, with the most important primary sources listed first, followed by translations that I have found useful and a few modern research directions where they seem relevant.

Where I have cited modern scientific research in the main text, I have done so with care and have generally pointed to the kind of research rather than to specific studies — partly because the underlying research is itself an active and evolving field, and partly because a printed book has limited usefulness as a citation manager for studies that may be superseded by the time the book is read. For specific studies, search the relevant scientific databases (PubMed for medical and biological research, JSTOR for the humanities, Google Scholar as a starting point for almost anything) using the keywords from the relevant chapter.

For the traditional Indian sources, I have listed the principal text and given the most readily available English translations. Many of these texts also exist in numerous editions in Sanskrit and Hindi and in the regional languages of India. The reader who can access them in the original languages will find dimensions that no translation can fully carry.

*A general note on translations: every translation is also an interpretation. Where possible, read more than one translation of a text. Where a phrase strikes you as particularly important, look at how different translators have rendered it. The act of comparison often reveals layers of meaning that any single translation must compress.*

## The Landscape of the Texts

For the reader meeting this literature for the first time, a one-paragraph map. The tradition sorts its texts onto two shelves. The first is Shruti — "that which was heard" — the four Vedas (Rig, Sama, Yajur, Atharva), each containing hymns, ritual manuals, forest treatises, and, as their philosophical conclusion, the Upanishads. This book's opening shloka, the Aham Brahmasmi of Chapter Eight, and the Shanti Mantra of Chapter Thirteen all come from this shelf. The second is Smriti — "that which was remembered" — everything composed by named or unnamed human authors to carry the Shruti's meaning into lived life: the two epics (the Ramayana and the Mahabharata, which contains the Bhagavad Gita), the eighteen Mahapuranas with their stories of Samudra Manthan and Dattatreya, and the practical sciences — Ayurveda's Samhitas, the Yoga Sutras, the Shilpa and Dharma Shastras. The full library is far vaster than this paragraph — one hundred and eight Upanishads by the traditional count, thirty-six Puranas, six schools of philosophy, and sciences from grammar to architecture — but the reader does not need the whole library to begin; the texts cited in these notes are the doors this book actually opened. The pattern this book has described — deepest knowledge in the smallest circle, stories for every household — is visible in the architecture itself: Shruti was guarded; Smriti was given to everyone.

## Foundational Texts Referenced Throughout

*These texts are referenced in multiple chapters; the reader will encounter them across the whole arc of the book.*

## Ashtanga Hridayam (Heart of the Eight Branches of Medicine)

Composed by Vagbhata, approximately the sixth or seventh century CE. One of the three foundational classical texts of Ayurveda (alongside the Charaka Samhita and the Sushruta Samhita). The first verse — the salutation to the physician who removed Raag, Dvesha and Moha without residue — frames the entire argument of this book.

Murthy, K. R. Srikantha (translator). Ashtanga Hridayam of Vagbhata. Three volumes. Krishnadas Academy / Chowkhamba Krishnadas Academy, Varanasi.

Hilgenberg, Luise and Kirfel, Willibald (German translation). Vagbhata's Ashtangahrdayasamhita. E.J. Brill, Leiden, 1941. A scholarly reference edition.

## The Upanishads

The philosophical concluding portions of the Vedas. The Isha Upanishad — whose opening Shanti Mantra (peace invocation) is the verse that opens this book and to which the Conclusion returns — is one of the shortest and most concentrated. The Brihadaranyaka, Chandogya, Katha, Mundaka, and Mandukya Upanishads carry the philosophical claims (Aham Brahmasmi, Tat Tvam Asi, the four states of consciousness) that Chapter Eight returns to.

Olivelle, Patrick (translator). The Early Upanishads: Annotated Text and Translation. Oxford University Press, 1998. The most rigorous modern scholarly translation in English.

Easwaran, Eknath (translator). The Upanishads. Nilgiri Press. A more accessible translation for the general reader.

Radhakrishnan, S. (translator and editor). The Principal Upanishads. HarperCollins. A philosophical edition with extensive commentary.

## The Bhagavad Gita

The dialogue between Krishna and Arjuna on the battlefield of Kurukshetra, embedded within the larger Mahabharata. Referenced in Chapters Four, Eight, Nine and Eleven. The second chapter, on the nature of the Atman, contains the verses that open Chapter Eleven.

Easwaran, Eknath. The Bhagavad Gita. Nilgiri Press. Widely used and accessible.

Sargeant, Winthrop (translator). The Bhagavad Gita. State University of New York Press. Word-by-word grammatical translation, useful for those wanting precision.

Radhakrishnan, S. (translator). The Bhagavadgita. HarperCollins. Translation with extensive philosophical commentary.

## The Ramayana and the Mahabharata

The two great Indian epics, both referenced throughout this book. The Ramayana of Valmiki is the original Sanskrit version; the Ramcharitmanas of Tulsidas, composed in the sixteenth century in Awadhi (a regional language of north India), is the version through which most Indian readers have encountered the story. The Mahabharata is much longer and contains the Bhagavad Gita.

Goldman, Robert P. and Sutherland Goldman, Sally (translators). The Ramayana of Valmiki. Seven volumes. Princeton University Press. The standard modern scholarly translation.

Hill, W. Douglas P. (translator). The Holy Lake of the Acts of Rama (Ramcharitmanas of Tulsidas). Oxford University Press, 1952. A serviceable English translation.

van Buitenen, J.A.B. and Fitzgerald, James L. (translators). The Mahabharata. University of Chicago Press (multi-volume, ongoing). The standard modern scholarly translation in English.

Ganguli, Kisari Mohan (translator). The Mahabharata. A nineteenth-century English translation available in full in the public domain online; the most complete single-translator English edition.

## The Puranas

The Mahapuranas (eighteen major encyclopedic texts) and Upapuranas (eighteen secondary ones) are the vast body of Indian narrative literature within which Samudra Manthan (Chapter Four), Dattatreya's twenty-four teachers (Chapter Fourteen), and many of the other stories in this book are preserved.

Bhagavata Purana — the source for Dattatreya's discourse in Canto Eleven. Available in many translations; Swami Prabhupada's edition is widely circulated.

Vishnu Purana — one of the earliest sources for Samudra Manthan. Wilson's nineteenth-century translation is available in the public domain.

Agni Purana — referenced in Chapter Seven as a source on the creation of Mantras and the technical specifications of ritual. The English translation by M.N. Dutt (1903) is in the public domain.

## Chapter One — Raag: The Disease Called Desire

The foundation of the chapter is the first verse of the Ashtanga Hridayam (listed in the foundational texts above). The reader who wishes to investigate Raag, Dvesha and Moha in depth should turn to that verse and its traditional commentaries.

The chapter also references several core concepts from Patanjali's Yoga Sutras — particularly the Kleshas (afflictions): Avidya, Asmita, Raga, Dvesha, and Abhinivesha.

Iyengar, B.K.S. Light on the Yoga Sutras of Patanjali. HarperCollins, 1993.

Bryant, Edwin F. (translator). The Yoga Sutras of Patanjali. North Point Press, 2009. A modern scholarly translation with extensive commentary.

*On the science of habituation and the brain's response to repeated stimuli — the modern scientific frame within which the Indian psychology of Raag can also be read — readers may begin with research on dopamine and the reward system, hedonic adaptation, and Sheena Iyengar's work on the paradox of choice.*

## Chapter Two — White: The Ancient Science of Equality

The traditional sources for the discussion of Vasudhaiva Kutumbakam are the Maha Upanishad and the Hitopadesha (Friendly Counsel, a classical collection of moral fables).

For the discussion of color in Indian ritual — the red, yellow, black, green, and saffron linens for different purposes — the principal reference is the Tantric ritual literature and the regional ritual manuals (Paddhati texts) that have been transmitted within specific lineages. These are not always easily available in English translation, but introductions can be found in:

Padoux, André. The Heart of the Yogini: The Yoginihrdaya, a Sanskrit Tantric Treatise. Oxford University Press, 2013.

Flood, Gavin. The Tantric Body: The Secret Tradition of Hindu Religion. I.B. Tauris, 2006.

*On the science of color and visual uniformity in social settings, the field of color psychology is genuinely contested. The careful reader will find that strong claims often outrun the evidence. The chapter has tried to reflect this.*

## Chapter Three — Mahalakshmi: The Physics of Prosperity

The traditional account of Mahalakshmi in her three forms (Sampatti, Siddhi, Nidhi) is found across the Devi Mahatmya (Greatness of the Goddess) within the Markandeya Purana, the Lakshmi Tantra, and the various Stotras (hymns of praise) of Lakshmi.

Coburn, Thomas B. (translator). Encountering the Goddess: A Translation of the Devi-Mahatmya and a Study of Its Interpretation. SUNY Press, 1991.

Gupta, Sanjukta (translator). Lakshmi Tantra: A Pancaratra Text. E.J. Brill, 1972.

On Padmaja (Lakshmi seated on the lotus) and the lotus symbolism, the principal sources are the Sri Sukta within the Rig Veda Khila portion and the iconographic descriptions in the various Shilpa Shastras (treatises on art and craft).

*On the Hormones River section: introductory readings in endocrinology, the gut-brain axis, and the body's stress response systems may be found in any contemporary university-level physiology textbook. For the lay reader, popular works by Robert Sapolsky on stress, and by Emeran Mayer on the gut-brain connection, offer accessible starting points.*

## Chapter Four — Samudra Manthan: The Original Warning

The story of Samudra Manthan appears in multiple sources with some variations. The most complete versions are in the Vishnu Purana (Book One), the Bhagavata Purana (Canto Eight), and the Mahabharata (Adi Parva).

For the Marma (vital points) discussion that connects Samudra Manthan to the body's healing system, the principal text is the Sushruta Samhita.

Bhishagratna, Kunja Lal (translator). Sushruta Samhita. Three volumes. Available in the public domain through the Internet Archive.

Frawley, David, Ranade, Subhash, and Lele, Avinash. Ayurveda and Marma Therapy: Energy Points in Yogic Healing. Lotus Press, 2003.

## Chapter Five — The Cow and The Universe

The Kamadhenu (wish-fulfilling cow) appears in numerous Puranic sources. Krishna's identification with cowherding is central to the Bhagavata Purana (especially Cantos Ten and Eleven).

For the modern scientific research on cow products — including studies on milk, ghee, and traditional uses of cow dung as an antimicrobial floor coating — readers should search the relevant Indian science research databases (the Council of Scientific and Industrial Research, the Indian Council of Medical Research). Be aware that some of this research is contested; the chapter has tried to reflect what has reasonable evidence behind it and to leave aside the more sensational claims.

## Chapter Six — The Electricity Within You

On the bioelectric systems of the human body, modern textbooks of physiology and neuroscience are the standard reference. For the lay reader, a few accessible starting points:

Becker, Robert O. and Selden, Gary. The Body Electric: Electromagnetism and the Foundation of Life. William Morrow, 1985. An older but still useful introduction to the bioelectric perspective on the human body.

Oschman, James L. Energy Medicine: The Scientific Basis. Churchill Livingstone, 2000.

On the consume-excrete cycle and ecological balance, the foundational modern work is in systems ecology. Howard T. Odum's work, and the more accessible introductions by Fritjof Capra (especially The Web of Life), are starting points.

For Pranayama and the breath sciences of the Yoga tradition, see Patanjali's Yoga Sutras (cited above) and the Hatha Yoga Pradipika (Light on Hatha Yoga) by Svatmarama, fifteenth century.

Akers, Brian Dana (translator). The Hatha Yoga Pradipika. YogaVidya.com, 2002.

## Chapter Seven — The Inventions You Cannot See

On the Mantra creation method and the Agni Puran's technical specifications:

Dutt, Manmatha Nath (translator). Agni Puranam. Two volumes. Published 1903, available in the public domain through the Internet Archive.

On the Yantra tradition and the geometry of consciousness:

Khanna, Madhu. Yantra: The Tantric Symbol of Cosmic Unity. Thames and Hudson, 1979.

Brooks, Douglas Renfrew. Auspicious Wisdom: The Texts and Traditions of Srividya Sakta Tantrism in South India. SUNY Press, 1992.

On Tantra as a science of the universal network (not the popular reduction to sexuality):

Flood, Gavin. The Tantric Body (cited above).

White, David Gordon. The Alchemical Body: Siddha Traditions in Medieval India. University of Chicago Press, 1996.

On Karmkand (ritual technology) and the Vedic ritual system:

Staal, Frits. Agni: The Vedic Ritual of the Fire Altar. Two volumes. Asian Humanities Press, 1983.

Witzel, Michael. Various academic papers on Vedic ritual and oral tradition, available through Harvard University's South Asian Studies repositories.

## Chapter Eight — The Receiver: What Is Already Within

The principal Upanishadic sources for Aham Brahmasmi and the universe-as-recording argument are the Brihadaranyaka Upanishad (especially 1.4.10), the Chandogya Upanishad (Tat Tvam Asi, 6.8.7), and the Mandukya Upanishad with Gaudapada's Karika.

Olivelle, The Early Upanishads (cited above).

Nikhilananda, Swami (translator). The Mandukya Upanishad with Gaudapada's Karika and Shankara's Commentary. Advaita Ashrama. Available in multiple editions.

On the modern physics that the chapter draws parallels with — the Holographic Principle, the Cosmic Microwave Background, the persistence of electromagnetic signals — the lay reader may begin with:

Susskind, Leonard. The World as a Hologram. Journal of Mathematical Physics, 1995.

Greene, Brian. The Hidden Reality: Parallel Universes and the Deep Laws of the Cosmos. Knopf, 2011.

*On the lemon experiment, the placebo response, and consciousness-physiology connections: introductory reading in psychoneuroimmunology and the work on the placebo response by Ted Kaptchuk and others at Harvard Medical School offers a starting point. These are active fields and the science continues to develop.*

## Chapter Nine — Born Complete

The Eklavya story is preserved in the Mahabharata. The Arjuna paralysis and Krishna's response constitute the dramatic opening of the Bhagavad Gita.

On the Sutika period (postpartum care), the principal sources are the Charaka Samhita (Sharirasthana — section on embryology and women's health), the Sushruta Samhita (cited above), and regional postpartum manuals.

Sharma, Priyavrat (translator). Caraka Samhita. Four volumes. Chaukhambha Orientalia, Varanasi.

*Modern postpartum research has, in the last two decades, increasingly confirmed the wisdom of the kind of intensive, supported, isolated recovery period that the Sutika tradition prescribes. Readers may search for research on postpartum depression prevention, the role of social support in maternal mental health, and the comparative outcomes of different cultural postpartum practices.*

## Chapter Ten — Life Within Life

On the Varna discussion: the guna-karma formulation is Bhagavad Gita 4.13. The accounts of Vishwamitra, Vyasa, and Satyakama Jabala (Chandogya Upanishad 4.4) are in the standard translations cited elsewhere in these notes. The chapter's battlefield scene is offered explicitly as a thought experiment, not as recovered history; readers should be aware that the historical relationship between the textual Varna model and the lived jati system is a large and contested scholarly field, and the chapter's reading is one reading among several.

On the Microbiome and the human-microbial relationship, the modern scientific literature has exploded in the last two decades. Starting points:

Yong, Ed. I Contain Multitudes: The Microbes Within Us and a Grander View of Life. Ecco, 2016. An accessible introduction.

Knight, Rob. Follow Your Gut: The Enormous Impact of Tiny Microbes. Simon and Schuster TED, 2015.

On the brain-as-pharmacy and Ayurvedic approaches to chronic conditions including paralysis, the principal Ayurvedic sources are the Charaka and Sushruta Samhitas (cited above), supplemented by traditional Panchakarma manuals.

On Navratri, seasonal practice, and the Ardhanarishvara theology, the principal sources are the Devi Mahatmya (cited above) and the various Stotras (hymns) addressed to Shiva and Parvati in their united form.

Goldberg, Ellen. The Lord Who Is Half Woman: Ardhanarishvara in Indian and Feminist Perspective. SUNY Press, 2002.

## Chapter Eleven — The Life Cycle

The principal sources for the doctrine of the Atman (soul) and its journey are the Upanishads (especially the Katha and the Brihadaranyaka), the Bhagavad Gita (Chapter Two), and the various Indian philosophical schools' treatments of consciousness and continuity.

On the academic research into past-life memory and near-death experiences, the field is active and contested. The Division of Perceptual Studies at the University of Virginia (formerly led by Dr. Ian Stevenson and now under Dr. Jim Tucker) has produced the largest body of methodologically rigorous research on these topics. Their published work — available through standard academic databases — is the best starting point for the reader who wishes to investigate this seriously.

*The chapter has deliberately framed this research in general terms. Readers seeking specific cases and methodological details should consult the published academic literature directly.*

## Chapter Twelve — The Mirror: What Indians Think

This chapter is drawn from personal conversation and observation rather than from textual sources. The answers recorded here are the answers people gave, in their own words and their own languages, across many years of asking.

## Chapter Thirteen — The Last Day

On the modern mnemonic system described: Lorayne, Harry, and Jerry Lucas. The Memory Book. Stein and Day, 1974. On the Katapayadi consonant-numeral system and its use by the Kerala school of astronomers, see Plofker, Mathematics in India (cited above).

On the oral transmission of the Vedas and the redundancy of the Patha (recitation) system, see the works on Vedic ritual by Frits Staal and Michael Witzel (cited above). The mathematical-computational analysis of the phonetic structure of Vedic Sanskrit is an active area of research; the reader will find papers on this through the standard linguistics journals.

On the history of zero, India's contribution to mathematics, and the philosophical concept of Shunya:

Joseph, George Gheverghese. The Crest of the Peacock: Non-European Roots of Mathematics. Princeton University Press, 2010.

Plofker, Kim. Mathematics in India. Princeton University Press, 2008.

Seife, Charles. Zero: The Biography of a Dangerous Idea. Penguin, 2000.

## Chapter Fourteen — Panchabhuta: The Blueprint for Future Invention

Dattatreya's discourse to King Yadu is in the eleventh canto of the Bhagavata Purana, also called the Uddhava Gita (the discourse to Uddhava).

Tapasyananda, Swami. The Uddhava Gita: The Last Message of Sri Krishna. Sri Ramakrishna Math, Chennai.

On the Panchabhuta (five elements) framework and its applications across Indian thought:

Taittiriya Upanishad — particularly the second valli (chapter) on the structure of creation through the elements. Olivelle's translation (cited above).

Lad, Vasant. Ayurveda: The Science of Self-Healing. Lotus Press, 1984.

On Shri Anna (millets) and the resurgent scientific interest in ancient Indian grains, the International Year of Millets (2023, declared by the United Nations on India's proposal) generated substantial documentation. The Indian Institute of Millets Research at Hyderabad is the principal scientific resource.

## A Final Word

Two final notes on the use of this section.

First — the absence of any source from this list does not imply the absence of authority behind a claim made in the book. Much of what this book has argued is rooted in oral tradition, in practices transmitted from grandmother to granddaughter, in observations made across the unbroken chain of generations that this book has tried to honour. Not all of this is in books. Some of the most important sources for the arguments in this book are people who have never written anything down.

Second — the reader who wants to investigate any of the claims in this book further is encouraged to do so. The book has not asked anyone to take its arguments on faith. The Indian tradition itself, when it was at its most rigorous, asked its students to test what they were taught — to verify it in their own experience, to find the version of the teaching that worked in their own life and discard what did not. That is also my hope for this book. Carry what is useful. Investigate what seems incomplete. Set aside what does not fit your own honest understanding. The arguments in these pages are mine. The truth they point toward does not belong to me, and I have no claim on what you do with it.

*Further sources, updates to this list, and links to specific studies and translations will be maintained at whyindians.com as the project continues.*

Made in United States
Troutdale, OR
11/03/2026

56845256R00118